Porcelain to Silica Bricks

Howell G. M. Edwards

Porcelain to Silica Bricks

The Extreme Ceramics of William Weston
Young (1776–1847)

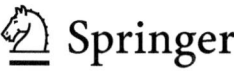 Springer

Howell G. M. Edwards
Department of Chemical and Forensic
Sciences, Faculty of Life Sciences
University of Bradford
Bradford, UK

ISBN 978-3-030-10572-3   ISBN 978-3-030-10573-0   (eBook)
https://doi.org/10.1007/978-3-030-10573-0

Library of Congress Control Number: 2018966382

A Nantgarw porcelain coffee cup, ca. 1820, locally decorated by Thomas Pardoe for the Spence-Thomas breakfast service during the tenure of William Weston Young as proprietor of the Nantgarw China Works. Glazed with the Nantgarw No. 2 glaze formulation of Young and Pardoe.

This Springer imprint is published by the registered company Springer Nature Switzerland AG
The registered company address is: Gewerbestrasse 11, 6330 Cham, Switzerland

Contents

About the Author

Prof. Howell G. M. Edwards M.A., B.Sc., D.Phil., C.Chem., FRSC is Professor Emeritus of molecular spectroscopy at the University of Bradford. He read chemistry at Jesus College of the University of Oxford and studied for his doctorate there before completing his B.A. and B.Sc. and taking up Research Fellowship at Jesus College, University of Cambridge. He joined the University of Bradford as Lecturer in structural and inorganic chemistry, became Head of the Department of Chemical and Forensic Sciences and was awarded Personal Chair in molecular spectroscopy in 1996. He has received several international awards in a distinguished spectroscopic career which has resulted in the publication of almost 1300 research papers in Raman spectroscopy and the characterisation of materials, along with six books on its application to art, archaeology and forensic analysis. He has had a lifelong interest in the porcelains of William Billingsley, especially those from the Derby, Nantgarw and Swansea china factories. He has authored two major books on Nantgarw and Swansea Porcelains: *Swansea and Nantgarw Porcelains: A Scientific Reappraisal* and *Nantgarw and Swansea Porcelains: An Analytical Perspective*, both published by Springer, Dordrecht. He has also produced several monographs on the personalities involved in these factories: *William Billingsley—The Enigmatic Porcelain Artist, Decorator and Manufacturer, Nantgarw Porcelain—The Pursuit of Perfection, Swansea Porcelain—the Duck-Egg Translucent Vision of Lewis Dillwyn* and *Derby Porcelain: The Golden Years, 1780–1830*. Howell Edwards is Honorary Scientific Adviser to the de Brecy Trust on the scientific evaluation of artworks and paintings.

List of Figures

List of Tables

Abstract

This book traces the life of William Weston Young, an entrepreneur, businessman and inventor, whose financial support was critically material in the establishment of the Nantgarw China Works in the early nineteenth century. Although the Nantgarw China Works is generally acknowledged as the brainchild of William Billingsley, ably assisted by Samuel Walker, who had the vision to create the world's most translucent and perfect soft paste porcelain as a canvas for his superb, decorative enamel painting, the influence of William Weston Young has often not been fully appreciated. After an unsuccessful attempt to start the Nantgarw China Works in 1814, the three went separate ways—Billingsley and Walker together joined Lewis Weston Dillwyn at the emergent Swansea China Works, where the fabulous and esteemed duck-egg porcelain paste was created, until in 1817, Billingsley, Walker and Young again teamed up to reopen the China Works again at Nantgarw, being entirely reliant for this enterprise upon the financial support from Young and his local business contacts. The new porcelain was an immediate success with purchasers in London society but became a victim of its own perfection in that supplies could not meet a seemingly inexhaustible demand from a discerning clientele, which resulted from an appallingly high kiln wastage approaching 90%. When only one item in ten in a commercial manufacturing process is deemed perfect and fit for sale, then it is surely impossible to recoup losses incurred in the production process, however good the product, so the Nantgarw China Works was forced to close again in 1820, having achieved Billingsley's original objective but failing dramatically to succeed as a commercial venture. Despite this setback, William Weston Young shouldered the sole responsibility for the ongoing operation of the Nantgarw China Works and the disposal of its assets in 1820 following the departure of William Billingsley and Samuel Walker for their employment at the Coalport China Works. The story might have ended there were it not for Young's invention of the high-temperature, *Dinas* refractory silica firebrick, whose incorporation into kiln and furnace linings facilitated the smelting of iron ore and copper ore at elevated temperatures, and also in glass manufacture, in industrialised nations through the nineteenth century. Hence, the title of this book reflects the vital contributions made by William Weston Young to these two extremes of ceramic technology: the first of these being an awesomely beautiful ceramic artwork which today, some two hundred years later, justly still retains its accolade as the *world's finest porcelain*, and the second the humble *Dinas* refractory silica firebrick, without which the heavy metal industries fuelling the growth of industrialised nations through the 1800s would not have been possible. The interactions between Young and key personalities of his acquaintance, and how these assisted in the development and realisation of these two extreme ceramic technologies, are also explored in depth. Finally, the evaluation of documentation which confirms that Young did in fact also seek to continue the manufacture of porcelain at the Nantgarw China Works in the years 1820–1823 and, indeed for some years thereafter, has been correlated with the recent chemical analysis of porcelain shards excavated from the waste pit at the Nantgarw China Works which affirms that he could have succeeded in making a hard paste porcelain there, contrary to statements in the historical literature that porcelain manufacture at Nantgarw ceased with the departure of Billingsley and Walker for the Coalport China Works in 1820. The holistic forensic approach involving the correlation between scientific analysis and historical documentation has enabled the correction of this assumptive and fundamentally incorrect statement and has alerted the community to search for evidence extant for porcelain made and decorated by William Weston Young and Thomas Pardoe at Nantgarw initially between 1820 and 1823, and then again also by Young and his associates in the period between 1830 and 1845.

Chapter 1
Introduction

Abstract The life and times of the late 18th and early 19th Century into which William Weston Young was born. Review of earlier literature and setting out the scope of this book with the two main themes of fine porcelain and the refractory silica brick.

Keywords Porcelain · Silica brick · Refractory · Nantgarw · Ceramic · Dinas · Industrial revolution

The title of this book reflects two extremes of ceramic technology and application: from the aesthetically pleasing and luxurious art form epitomised by highly translucent and delicate porcelain beautifully decorated with coloured enamels (Fig. 1.1) to the purely functional but visually much less attractive silica refractory bricks used as linings to the furnaces employed in heavy industry (Fig. 1.2) for iron and steel smelting and glass manufacture. In the heyday of the industrial revolution in Europe the profits made from the production of the latter were invariably enjoyed by investment and sponsorship by clientele of the former: the result was that out of the dirt and grime associated with the 19th Century heavy industry factories came forth some of the most beautiful and delicate ceramic artwork ever created, which even today is highly appreciated by connoisseurs of fine porcelain and the *cognoscenti*.

William Weston Young holds a niche position in this context as he was responsible with other key personnel for the manufacture of some of the world's finest and most translucent porcelain at the Nantgarw China Works in South Wales between 1817 and 1820 and also for the world famous refractory *Dinas* silica bricks between 1822 and 1828, without which the high temperature chemical reduction of iron ore in the blast furnaces and the smelting of copper ores in the reverberatory furnaces of highly industrialised countries would have been achieved with only great difficulty in the 19th Century. Both of these highly valued ceramics originated in their invention and manufacture in two valleys in South Wales, namely the Vale of Neath for the *Dinas* refractory silica firebrick in West Glamorgan and the Taff Vale for Nantgarw porcelain in East Glamorgan. Whereas the presence of William Weston Young in the manufacture of the finest porcelain at Nantgarw has been cited in historical texts along with his colleagues and partners, William Billingsley and Samuel Walker, his perceived minor role in this endeavour has been minimised relative to their own

© Springer Nature Switzerland AG 2019
H. G. M. Edwards, *Porcelain to Silica Bricks*,
https://doi.org/10.1007/978-3-030-10573-0_1

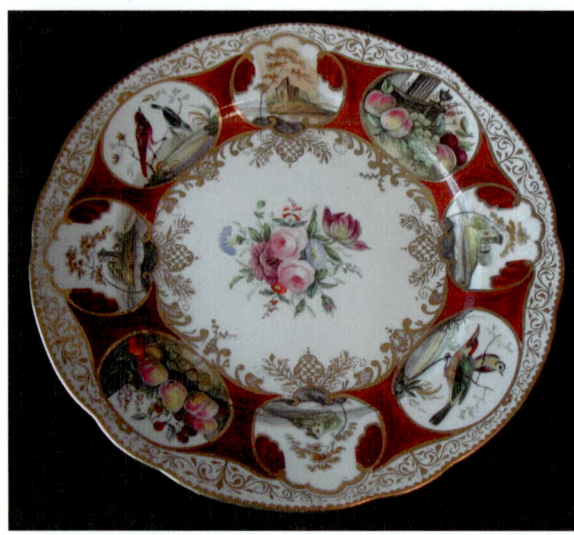

Fig. 1.1 Nantgarw porcelain, 1818: dinner plate from the *Duke of Cambridge* dinner-dessert service presented to Adolphus, Duke of Cambridge, by his eldest brother, George, Prince Regent and later King George IV, on the occasion of his marriage to Princess Augusta Wilhelmina Louise, Landgravine of Hesse-Kassel, June 1818, in Buckingham Palace. Decorated with fruit, flowers, birds and landscapes in vignettes on a claret ground by James Plant in the atelier of Robins and Randall for John Mortlock of Oxford Street, London. Impressed mark: NANT-GARW C.W. The surviving items of this service were dispersed at the auction of the late Duchess of Cambridge's effects in London in 1889. Private collection

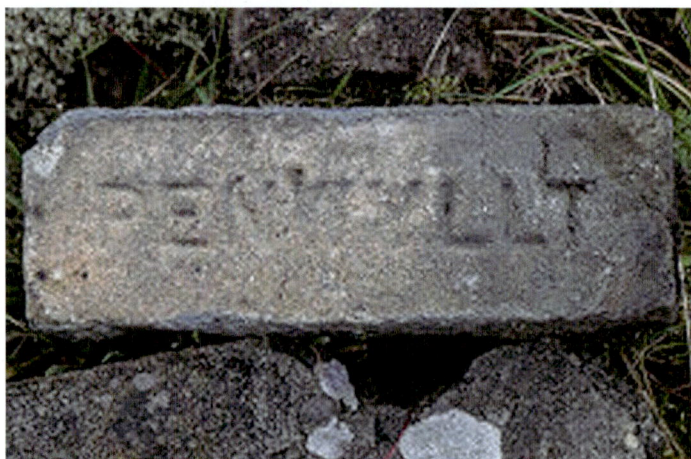

Fig. 1.2 A silica brick from the Penwyllt factory, Craig-y-Nos, Swansea Valley, ca. 1830, impressed PENWYLLT, in Welsh, meaning "*Wild headland*"

influence and his own contribution to industrialised society through his invention of the refractory silica brick has also been rather ignored.

In this book, the life of William Weston Young is traced from his early Quaker background through his involvement in a host of apparently unrelated polymathic activities and professions, to his lifelong interest in the decoration of ceramics and their manufacture, including his invention of the refractory silica brick. His dogged determination to succeed even after suffering two business bankruptcies along the way will be surveyed along with his interactions with people in his circle who played key roles in his efforts to reach his objectives. It is also interesting that Young's practical realisation of the two ceramic extremes illustrated here was actually achieved within the narrow temporal range centred on the year 1820 ± 3 years, and it must therefore be considered that his earlier delivery on the porcelain would therefore have contributed significantly to the creation of the silica brick: from historical research, it is clear that in the middle of this time frame, i.e. 1820–1821, he must have been undertaking synchronous trial experiments on both porcelain and bricks—and that he probably, in fact, initially used the same kilns at the Nantgarw China Works for his porcelain and brick trials to achieve this purpose.

This book could have been assigned a specific sub-title of "*The Rise and Fall of the Swansea and Nantgarw Porcelain Factories*," since both institutions were linked to William Weston Young and to several of the persons nominated therein: another alternative might have been "*Serendipity in Porcelain Manufacture in Georgian Britain*", which reflects the fortunes and misfortunes associated with the manufacture of this mysterious material carried out in the most empirical way, long before the scientific basis and understanding of the high temperature chemistry underpinning the successful creation of these beautiful and utilitarian works of art was appreciated. These titles ignore, of course, the contributions made to an industrialised society by William Weston Young in the synthesis of his refractory silica bricks and it will be a major objective in this book to bring the two extremes of the ceramic world together and to hypothesise that the empirical experimentation that was carried out on the one in fact assisted materially in the successful production of the other, perhaps in a way that has not been appreciated hitherto in the literature. The fact that the chemical composition of his refractory 19th Century *Dinas* silica bricks could be envisaged as being perhaps similar in some respects to that of the finest Chinese hard paste porcelains would not have been lost upon the entrepreneurial Young, who according to his diaries was trying to lift the Nantgarw China Works from its demise after the departure of William Billingsley and Samuel Walker by empirical experimentation on the manufacture of a high silica hard paste porcelain variety there at the same time as he was developing his *Dinas* high silica content refractory bricks.

The wonderful translucency of the porcelain manufactured at Swansea and Nantgarw over a few years only in the second decade of the nineteenth century (as demonstrated in the Nantgarw spill vase shown in Fig. 1.3, photographed in transmitted light to illustrate the perfection of the translucency) is universally credited to the activities of William Billingsley, who set out to achieve the creation of the best porcelain ever made to act as a canvas for his exquisite ceramic artistry in flower painting—a flawless, smooth textured, white soft paste body with crisp, moulded features and a soft

Fig. 1.3 Nantgarw porcelain, ca. 1817–1819, cylindrical spill vase decorated in London with dentil edge gilding and sprays of enamelled garden flowers. The photograph has been taken in transmitted light to illustrate the superb translucency and perfection of the porcelain, which is blemish-free and of exceptional clarity. Private collection

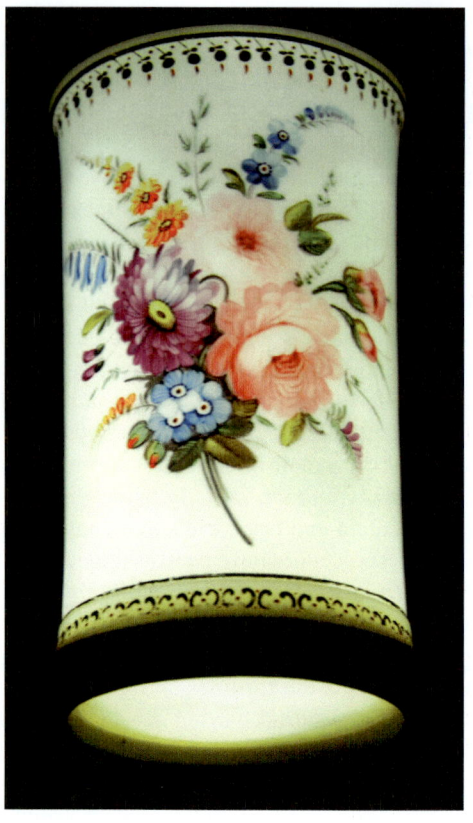

white glaze that enhanced the depth and perfection of the applied decorative enamels. However, although the initiation of the Nantgarw China Works project and the motivation and driving ambition to see it through in extreme adversity is undoubtedly fairly attributed to William Billingsley historically, it is equally clear that he could not have achieved it alone: his interactions with Lewis Dillwyn, Samuel Walker and William Weston Young were critical to the eventual successful achievement of his ambition to create the world's finest and most translucent soft paste porcelain. It will transpire that William Billingsley could never have achieved his porcelain perfection at the Nantgarw China Works without the material support and financial generosity of William Weston Young. Billingsley eventually achieved his lifelong ambition from some initial early and rather clandestine trials performed during his apprenticeship to William Duesbury I, at the Derby China Works, through his first embarkation into commercial porcelain manufacture with John Coke at Pinxton, to his subsequent attempts at Mansfield and Brampton-in-Torksey, where he first met Samuel Walker and welcomed him into his family as the husband of his elder daughter, Sarah. Then, Billingsley and Walker together received employment at the Royal Worcester China Works headed by Martin Barr during the Barr, Flight and Barr period from 1808 until 1813, when Martin Barr, their major sponsor and supporter, died. The departure of Billingsley and Walker from Brampton-in-Torksey for Worcester has been a subject of some chronological debate, but it is noted that Sarah Billingsley, who decorated porcelain at Brampton-in-Torksey, participated as a witness in a marriage there in 1808, so it is thought that they must have left shortly afterwards. Billingsley and Walker left the Royal Worcester China Works to set up the Nantgarw China Works with William Weston Young in 1813. At first this was an unsuccessful venture, then meeting firstly with Lewis Weston Dillwyn at the Swansea China Works and then William Weston Young again at the re-start of the Nantgarw China Works, before finally Billingsley and Walker departed Nantgarw together to join John Rose at the Coalport China Works in 1820, where William Billingsley died on January 16th, 1828. Billingsley had already achieved his lifelong objective by this time, with Nantgarw china being accorded the esteemed and deserved accolade of "the world's finest porcelain" even though it was commercially not a viable commodity for sale despite its being held by a discerning clientele as the epitome of its type.

Jewitt (*Ceramic Art in Great Britain* 1878, p. 160) alleges that, following the arrival of Billingsley and Walker at Coalport in 1820, the Coalport body immediately improved in excellence and quality: however, Nance (*The Pottery and Porcelain of Swansea and Nantgarw* 1942, p. 38) has indicated that, despite any improvements that may have been made to the paste by Billingsley and Walker at Coalport, the 1820s Coalport body in his opinion was still harder and was rather unattractively a "cold white" in appearance, lacking the mellow richness of the analogous Nantgarw specimens. At this time, John Rose of the Coalport China Works patented his renowned lead-free glaze in response to the public concern being expressed nationally about the exposure of a young workforce in china factories to toxic and potentially lethal heavy metal contamination by skin absorption, especially the lead oxide in glazes and metallic mercury in gilding amalgams. A special comment made by James Rouse, a decorator at Coalport, confirms that the Coalport body improved after the arrival there of Billingsley and Walker in 1820 and he even suggests that this new porcelain body was moulded and decorated in the Nantgarw style and was "*sold as such*" (Jewitt, *Ceramic Art in Great Britain* 1878, p. 160)—even then, it seems that industrial espionage and sharp marketing practice was being utilised to increase the potential sales of a product! Indeed, the strong implication is that a conscious effort was made at the Coalport China Works to simulate and possibly even recreate Nantgarw china, which was produced and decorated at Coalport, to effectively supply a defined market need and shortage that was already apparent in the 1820s. The precise terms of employment of both Billingsley and Walker at Coalport are unknown, but Dillon (*Porcelain and How to Collect It* 1910, p. 287) is of the opinion that Billingsley continued to paint at the Coalport China Works, even though Dr. John asserts that his role was more likely to have been in an advisory, instructing and supervisory capacity in the Coalport enamelling workshop (John, *Nantgarw Porcelain* 1948, *William Billingsley* 1968). Samuel Walker, however, according to Jewitt (*Ceramic Art in Great Britain* 1878), was employed in the paint and glaze mixing room at the Coalport China Works and was described by him as a "*very clever workman*", doing much to improve the quality of the output from the workshop and that he was especially in demand for advising on the engineering aspects of the art of china manufacture. For example, it is on record in the *Pottery Gazette* of 1885, that Walker in 1821 introduced a very special maroon ground colour and associated glaze at the Coalport China Works which comprised: "*A pink maroon glaze, 14 parts stone, 12 parts borax, 2 parts whiting, 2 parts lead oxide, 4 parts tin ash and 5 parts crimson oxide*". Apparently, this particular decoration was greatly appreciated by the discerning Coalport china clientele.

Despite their early historical accounts, there are several inconsistencies to be noted in both the writings of Jewitt (*Ceramic Art in Great Britain* 1878) and Haslem (*The Old Derby China Factory* 1876) relating to Nantgarw porcelain manufacture and its associated personalities, one of which is the recording of Samuel Walker as "*George Walker*", which arose from these authors mis-interpreting the Nantgarw China Works' impressed mark of NANT-GARW C.W. (Fig. 1.4) as NANT-GARW G.W—where the letters G.W. were unrealistically and incorrectly assumed by Jewitt and Haslem to be the initials of George Walker! Another misinterpretation of the underlying meaning associated with this Nantgarw impressed mark historically has been that the letters C.W. incorrectly stood for Cardiff, Wales and not China Works!

Our porcelain theme, which centres on William Weston Young, therefore, features necessarily the interlinked careers of Dillwyn, Billingsley, Pardoe and Walker and how these impinged upon the activities of Young by placing their contribution to porcelain manufacture in its properly apportioned context along with Young's many

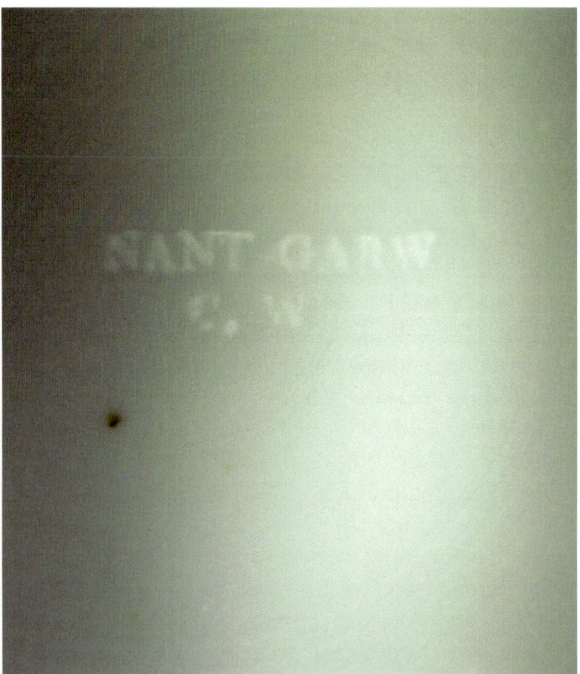

Fig. 1.4 The Nantgarw impressed mark, NANT-GARW C.W., used on flatwares and best seen in transmitted light as shown in this photograph of the mark on a Nantgarw plate from the *Phippes* armorial service, ca. 1817–1819. Private collection

other ventures and activities which have not always been addressed fully in context hitherto. A similar exercise will be undertaken for the creation and manufacture of the *Dinas* silica refractory brick after experimentation by Young, for which it appears he generated the idea, sourced the components and executed the synthesis alone, the product of which, like Nantgarw porcelain, far outstripped its erstwhile competitors internationally and, moreover, held a prime niche position for many years. However, both of these ventures, although successful in their eventual objectives, did not make Young a wealthy man—indeed, far from it—and he died in financially restricted circumstances, being a victim of unfavourable local and global events and more than his share of accumulated bad fortune and mismanagement in his business dealings. Turner (*The Ceramics of Swansea and Nantgarw* 1897) has expressed his opinion of William Weston Young as "*An inventor who certainly did a good thing for our manufacturers even if he did not benefit much himself personally from their production*".

To place the scene in its proper historical context, it should be appreciated that the accelerated growth of a heavily industrialised Great Britain in the early 1800s afforded trade and manufacturing opportunities which were offset by burgeoning unrest caused by a series of colonial conflicts and global hostilities which were a constant drain on the nation's wealth and resources. William Weston Young was born in 1776, when the secessionist states of America declared their official independence from Great Britain; the French Revolution started in 1789 with the storming of the Bastille, and this generated the rise of Napoleon and the Napoleonic Wars which lasted until 1815, followed by a series of more localised engagements such as the war between Great Britain and the emergent United States of America in 1812 and the uprisings against British rule in colonial India and Africa in the early to mid-19th Century, such as the Mahratta Wars and the First Zulu War. In particular, these events restricted trade opportunities and access to raw materials, especially during the first two decades of the 19th Century, which saw the growth and then the rapid subsequent closure of the Swansea and Nantgarw porcelain manufactories, along with several others. It was as a direct result of these scenarios involving international unrest that William Weston Young was forced into his first bankruptcy as a "corn factor" and saw his resultant move from a strictly local agricultural base of business operations initially into ceramics decoration in association with Lewis Weston Dillwyn at Swansea, where they also shared an interest in local natural history and in earthenware china painting. This latter activity surely nurtured Young's interest in china, which occupied his talents over the next twenty years and for some while thereafter, sporadically, almost up to the time of his death in 1847.

The format of this text will comprise a description of the life and aspirations of William Weston Young and how these impacted upon his local community, and especially highlighting his interactions with the other personalities exemplified here who played such a critical role in the achievement of his scientific and ceramic objectives, particularly Lewis Weston Dillwyn, William Billingsley, Thomas Pardoe and Samuel Walker. Outside of this, the broad interests of William Weston Young in natural history, art, botany and zoology, entomology, geology, wreck-recovery, land surveying and general agricultural merchandise production such as flour and corn milling will be explored. The seminal work of Ernest Morton Nance, entitled "*The Pottery and Porcelain of Swansea and Nantgarw*" (1942), although not the first to be written specifically on the factories of Swansea and Nantgarw after some 40 intensive years of his own research, did manage to correct several opinions stated by William Turner some fifty years before in his "*Ceramics of Swansea and Nantgarw*", published in 1897, the first book dedicated to the two Welsh ceramics factories. Until that time, the comments of Turner had been treated with the greatest respect and unarguable credibility as being absolutely authoritative on account of the personal interviews which he conducted with personnel such as Henry Morris and Richard Millward, who had actually worked as young men at the china factories of Swansea and Nantgarw, respectively, and who presumably would have provided him with apparently indisputable factual information about the people employed there and details of the manufacturing processes that were being undertaken at their time. However, as can be well appreciated nowadays, eye-witness statements of observers, however objectively and assuredly given, can frequently be misinterpreted forensically or unduly credited and emphasised, especially since Turner's information from his designated interviewees would necessarily have been derived from their recall of events that had occurred at least 60 years previously, as both the Swansea and Nantgarw china factories had closed down their porcelain production in 1820. In the mid-20th Century, several cognitive and authoritative texts appeared which further added to the literary corpus of the Swansea and Nantgarw factories and corrected other errors which had been perpetuated from the earlier works: these include the four books by Dr. William John and co-authors, namely "*Nantgarw Porcelain*", "*Swansea Porcelain*", "*William Billingsley*", and "*The Nantgarw Porcelain Album*" (with K. Coombes and G.J. Coombes), the text on "*Swansea Porcelain*" by A. "Jimmy" Jones and Sir Leslie Joseph and that on "*Swansea Porcelain*" by Elis Jenkins. The most recent books on these factories are those by the present author on "*Swansea and Nantgarw Porcelain: A Scientific Reapp*raisal", published in 2017, and "*Nantgarw and Swansea Porcelain; An Analytical Perspective*", which was published in 2018. Fergus Gambon has also published a monograph recently on the John Andrews collection of Welsh porcelain housed at Plas Glyn-y-Weddw, near Abersoch, in Gwynedd, North Wales (Gambon 2017). Several short monographs on personalities associated with the Swansea and Nantgarw factories by the present author have also appeared recently, namely "*William Billingsley: Porcelain Artist, Decorator and Manufacturer*", "*Duck-Egg Porcelain: The Translucent Vision of Lewis Weston Dillwyn*", and "*Nantgarw Porcelain: The Pursuit of Perfection*". In addition, several earlier authors have referred to opinions expressed by writers of more general texts on porcelains and ceramics such as Church (*English Porcelain* 1894), Chaffers (*Marks and Monograms on Pottery and Porcelain with Historical Notes on Each Manufactory* 1863), Jewitt (*Ceramic Art in Great Britain* 1878), Shaw (*The Chemistry of the Several Natural and Artificial Heterogeneous Compounds Used in the Manufacture of Pottery, Porcelain and Glass* 1837) and Haslem (*The Old Derby China Factory* 1876) who have included comments on the Swansea and Nantgarw factories in their content and have mentioned interesting notes on the personnel who worked there. In many cases, as indicated above, the information contained in these texts should advisedly be treated with caution, despite their being written in an authoritative and a credibly factual fashion, as in many cases the comments have been made from hearsay, albeit from personal interviews with people who actually worked at the factories as quoted in Turner (*The Ceramics of Swansea and Nantgarw* 1897), such as Richard Milward at the Nantgarw China Works and Henry Morris at the Swansea China Works, which are perhaps now worthy of re-consideration when subjected to independent verification against extant historical documentation.

Examples of such erroneous but initially plausible statements which appeared in these early writings are: that Samuel Walker was Billingsley's "*chemist*" (he was, in fact a highly reputable furnace and kiln engineer—he even described himself personally as a "mechanic" in some official documentation, but certainly not a chemist!); that Lewis Dillwyn did not appreciate the science and chemistry of his empirical porcelain body changes (yet he was an esteemed member of The Royal Society, to which he was elected in 1804 for his erudite scientific work (Dillwyn, 1804–1809) on British *Confervae* and freshwater algae, 1809), and it is clear that he certainly understood the basic chemistry of his processes from his notebook recipes and formulations, despite the contrary assertions suggesting otherwise of Shaw (Simeon Shaw, 1837); that Thomas Pardoe never painted at Nantgarw as he just was "not good enough" and all the local decoration therefore had therefore to be accomplished by William Billingsley (Robert Drane, writing in William Turner, *The Ceramics of Swansea and Nantgarw* 1897, who also believed that most Nantgarw china was sold locally and not through their London distributors or agents!); that William Weston Young bought in Swansea china from Lewis Dillwyn for decoration at Nantgarw and that John Rose bought all the hardware, moulds and furnace materials from Swansea and Nantgarw in 1820 upon closure of the factories, which he then re-erected and used at his Coalport China Works. Turner (*The Ceramics of Swansea and Nantgarw* 1897) himself comments that Pardoe had no originality and that he was characteristically a copyist, unlike William Billingsley, whose work was "*distinctively original and inspired by nature with a loving familiarity*"—which certainly does not reflect the opinion of modern aficionados of Pardoe's painting on Nantgarw porcelain!

The veracity of these and similar statements reflects materially upon the operations of the two china factories at Swansea and Nantgarw and has been examined and re-evaluated in the two recent publications by Edwards (*Swansea and Nantgarw Porcelains: A Scientific Reappraisal* 2017c, *Nantgarw and Swansea Porcelains: An Analytical Perspective* 2018). After 1818, Dillwyn ceased to be interested in the manufacture of porcelain at the Swansea China Works, which was then bought by Timothy and John Bevington, who ran it until the final closure of the Swansea China Works operation occurred in 1820. Dillwyn then seemed to concentrate his activities more in the local and national political arenas, becoming High Sheriff of Glamorgan, being elected a Member of Parliament for Glamorgan in 1834 and Mayor of Swansea in 1839—but he still retained a scientific interest in that he was instrumental in the formation of the Royal Institution in Swansea, and thence becoming its first President. Lewis Dillwyn died on the 31st August 1855, at his home in Sketty Hall, Swansea.

A recent authoritative publication by Ramsay and Ramsay (*The Evolution and Compositional Development of English porcelain from the 16th C to Lund's Bristol ca. 1750 and Worcester c. 1752–The Golden Chain* 2017) also comments on erroneous statements perpetuated by earlier authors relating to the porcelains of Bow and the mysterious "A-marked factory" which they have traced back to Chaffers (*Marks and Monograms on Pottery and Porcelain with Historical Notes on Each Manufactory* 1863) and which seemingly have not been checked against later evidential documentation. In a foreword to their seminal article, Ramsay and Ramsay quote Soame Jenyns (*Later Chinese Porcelain* 1971, p. 1), who cautions against the automatic acceptance of apparent "statements of fact" which have not been subjected to analysis and potential revision:

> But as the process of reassessment is always silently at work, time gradually imparts in turn the value of each standard book on ceramics. A shift of emphasis, a change of date or provenance in the light of new information may alter the whole surrounding scenery. In the world of scholarship to give and take criticism is all part of the day's work, and each of us in our turn may legitimately criticise our predecessors without being guilty of presumption, so long as we can look forward without rancour to being criticised in our turn by our successors, when our day is past. This is the inevitable destiny of all critics. But the author who is surpassed is not necessarily superseded. If the touchstone of his criticism proves his true metal he has added, like Hobson, one or more links to the golden chain, which long after his intervention is forgotten stands as his contribution to the subject.

In the current text, such unverified statements as alluded to here have been objectively considered specifically where they impinge upon the activities of William Weston Young and have been examined closely and questioned when found to be inconclusive or at variance with the established independent historical documentation, some of which, such as the *Diaries* of William Weston Young, listed in Table 3.1 (also referred to elsewhere as the *Fact-Books*), the *Diaries* of Lewis Weston Dillwyn, the *Notebooks* of Dillwyn (reproduced in Eccles and Rackham 1922, and in Appendix B) and the early Nantgarw china analyses of Eccles and Rackham (*Analysed Specimens of English Porcelain* 1922) were not available to the earliest researchers and commentators enumerated above, such as Chaffers (1863), Haslem (1876), Jewitt (1878), Church (1894) and Turner (1897). Firstly, we shall endeavour to set the scene by recounting relevant information about the early career and business ventures of William Weston Young and relating how his experiences with people he met developed into his business relationships and ceramics manufacture of his two extreme ceramics, namely the delicate and fragile Nantgarw china and the robust *Dinas* refractory silica bricks.

William Weston Young, has been described as "*gifted, and of great physical and mental energy*" (Nance, *The Pottery and Porcelain of Swansea and Nantgarw, Appendix VIII: William Weston Young, 1776–1847* 1942): Turner (*The Ceramics of Swansea and Nantgarw* 1897) derived from interviews with people who knew him that William Weston Young had enormous physical strength and was a hard worker, and that he was observed to lift the blacksmith's anvil at Nantgarw easily and that weighed 8 cwt, approximately 900 lbs (400 kilos)! Nevertheless, throughout his life he was hardly free from financial worries and even when he made money through his primary professional land surveying business he lacked the financial acumen to keep it. His attributes were numerous: an artist and tutor of distinction in engraving, drawing and painting, he taught drawing to pupils, for example, Miss

Williams, the daughter of the Headmaster of the prestigious Cowbridge School, which at this time was the premier Free School in the Principality, regularly sending its students to Oxford University, a writer of prose and verse, an antiquarian, a man of science particularly with natural history interests, an architect and an inventor, holding patents for several procedures, including a grab mechanism for the raising of wrecks and for a novel process for subsurface salt extraction. Professionally, he was a farmer, a miller, a wreck-raiser, a general merchant and a land and colliery surveyor: starting life as a farmer and agricultural corn factor in the Neath valley, he ended it as a colliery manager in Kidderminster, yet he is best remembered for his work on ceramics. The two examples, for which he is renowned and which form the topic of this book, spanning the spectrum of creative ceramic production in the early 19th Century, namely the manufacture of the finest quality porcelain and of silica refractory bricks, which are located at the extremes of ceramic technology—the one being incredibly artistic and delicate and the other robustly functional. However, both of these ceramics had their following in different walks of life, from Georgian social etiquette and convivial entertainment to the heavy industrial production of iron and copper from their ores in the furnaces that drove the advancement in technology of the Industrial Revolution.

The major sources for information about William Weston Young are those of Nance (*The Pottery and Porcelain of Swansea and Nantgarw, Appendix VIII A and B* 1942, pp. 488–540), Turner (*The Ceramics of Swansea and Nantgarw* 1897), Jenkins (*Transactions of the Newcomen Society* 1942, in Stewart Williams' *Glamorgan Historian* 1968, Vol. 5, pp. 61–107) and more recently, the excellent chapter by Renton (Chap. 7, *Thomas Pardoe and William Weston Young,* in *Welsh Ceramics in Context*, Volume I, ed. J. Gray 2005). Morton Nance reproduces Young's diaries and work books in his *Appendix B*, comprising 45 pages of closely spaced text enumerating details of his accounts, travels and details of his daily meetings, which is a prime source of information, if not very readable in context, reflecting the shorthand nature of the handwritten entries made by Young himself. In contrast, the account of Elis Jenkins, although giving a balanced view of Young's relationships, does not provide anything relating Young's experiments with porcelain or bricks, the subject of our text here. An important and early source of information about William Weston Young is William Turner's book, since he was able to interview personally Colonel J. W. Young of Preswylfa, Neath, who was the grandson of William Young's elder brother, Joseph, and who was for some time the managing director of the Vale of Neath Firebrick and Ceramic Co. Ltd., which manufactured the famed Dinas refractory silica firebrick. He had retained several letters, correspondence and documents relevant to his great uncle, William Weston Young in the family archive. The article by Andrew Renton is not merely an extension of the available literature, comprising 26 pages with an impressively supporting 134 references, but provides a superb recounting of the complex relationship between Thomas Pardoe and William Weston Young with a wealth of detail and written in a very readable format. In contrast, Morton Nance's *Appendix A*, gives an introduction to the life of William Weston Young in only six pages, which is very detailed but still readable, despite the prevalence of lengthy and somewhat distracting footnotes, as highlighted by Elis Jenkins in his own book chapter. It is interesting, nevertheless, that some episodes of Young's life have been interpreted with different nuances. For example, the capture of Young aboard a vessel bound for North America by a French privateer in the Channel approaches just after he left Bristol on the 4th August, 1794, as described briefly in Morton Nance, appears in more detail in Jenkins' account as a fifteen-strong French fleet taking an American vessel, the *"Severn"* bound for the USA from Bristol, south of Ireland to a port in northern France and Young's escape soon thereafter to a neighbouring American vessel, which unfortunately had also been commandeered by the French and was being used as a prison ship—resulting in his recapture and eventual escape after three month's further incarceration! This episode, although of brief duration as he was back in Bristol and married to Elizabeth Davies on April 14th, 1795, certainly illustrates Young's determination to carry through a project and his underlying daring and strength of will to succeed, which will characterise his future career exploits and business relationships. Elizabeth's diary, published by William Weston Young after her death in 1842 (Young 1843), provides much more detail about the human side of her husband: apparently, his French captors failed to issue their prisoners with utensils to eat their soup so Young fashioned spoons from wood to enable them all to survive. In his escape from the French, he swam several miles, avoiding picket boats and guards *en route* to his freedom. Although attributed by most observers to the uncertainty in markets for corn and cereals following the Peace of Amiens in 1802, another interpretation of the cause of Young's bankruptcy is offered by Jenkins (in Williams, *Glamorgan Historian* 1968) who ascribes this event to the collapse of his local business dealings with some rather unscrupulous people and Young's business naivety rather than to the bottoming-out of the international price for corn, which he had stockpiled at significantly higher prices and could not then sell on, due to the cessation of the hostilities with Napoleonic France.

The account of Young's life given in the article by Jenkins (1942), entitled *The Silica Brick and Its Inventor, William Weston Young*, which was published in the same year as the book by Nance (1942), provides an excellent pastiche of his life, especially the later period when he discovered the technology of processing the silica-rich Dinas rock into an efficient and usable silica refractory brick. The earlier features of William Weston Young's life are also to be found in the Diary of Elizabeth Young, which was published after his wife's death on 10th March, 1842 at the age of 77 by William Weston Young, firstly under the auspices of his friend and business associate John Wright in Bristol in 1842, and thereafter by Harvey and Darton in London in 1843: this was entitled *The Christian Experiences of Elizabeth Young, a Member of the Society of Friends, and Written by Herself*, this last phrase added by William Weston Young. The Diary, according to Jenkins (*Glamorgan Historian* 1968), contains edited extracts from Elizabeth's notes between March 1791, before her marriage to William, until shortly before her death in 1842, and details the hardships she faced with William over nearly fifty years of marriage under sometimes difficult circumstances, during which Elizabeth summarises their long life together very nicely as: *"Our vicissitudes have been many, our mercies still more"*.

References

W.B. Chaffers, *Marks and Monograms on Pottery and Porcelain with Historical Notes on Each Manufactory* (J. Davy & Sons, London, 1863) Kessinger Legacy Reprints (Kessinger Publishing, Whitefish, 2010)

Sir A.H. Church, *English Porcelain: A Handbook to the China Made in England During the 18th Century as Ilustrated by Specimens Chiefly in the National Collection, A South Kensington Museum Handbook* (Chapman & Hall Ltd., London, 1885 and 1894)

E. Dillon, *Porcelain and How to Collect It* (Methuen and Son, London, 1910)

L.W. Dillwyn, *British Conferva or Coloured Figures and Description of British Plants Referred by Botanists to the Genus Conferva* (W. Phillips. George Yard, Lombard St., London, 1809)

L.W. Dillwyn, *Notes on the Experimental Production of Swansea Porcelain Bodies and Glazes*. Made by Lewis Weston Dillwyn with Samuel Walker at the Swansea China Works Between 1815 and 1817. Presented to the Library of the Victoria & Albert Museum, South Kensington, London by John Campbell in 1920. Reproduced in Eccles & Rackham, *Analysed Specimens of English Porcelain*, 1922, see reference below

H. Eccles, B. Rackham, *Analysed Specimens of English Porcelain in the Victoria and Albert Museum* (Victoria and Albert Museum, London, 1922)

H.G.M. Edwards, *Nantgarw Porcelain: The Pursuit of Perfection,* Series Editor: M.D. Denyer (Penrose Antiques Ltd. Short Guides, Thornton, 2017a). ISBN 978-0-244-90654-2

H.G.M. Edwards, *Swansea Porcelain: The Duck-Egg Translucent Vision of Lewis Dillwyn* (Penrose Antiques Ltd. Short Guides, Thornton, 2017b). ISBN 9780244325787

H.G.M. Edwards, *Swansea and Nantgarw Porcelain: A Scientific Reappraisal* (Springer, Dordrecht, 2017c)

H.G.M. Edwards, *Nantgarw and Swansea Porcelain: An Analytical Perspective* (Springer, Dordrecht, 2018)

H.G.M. Edwards, M.C.T. Denyer, *William Billingsley The Enigmatic Porcelain Artist, Decorator and Manufacturer* (Penrose Antiques Ltd., Short Guides, Neopubli, Berlin, 2016). ISBN 978-3-7418-6802-3

F. Gambon, *Porslen Abertawe a Nantgarw/Swansea and Nantgarw Porcelain*, Catalogue of the F.E. Andrews Collection in the Oriel Gallery. Plas Glyn-y-Weddw, Pwllheli, Llanbedrog, Gwynedd, North Wales. John Andrews Charitable Trust Publishers/JACT Oriel Gallery, Llanbedrog, UK, 2017

J. Haslem, *The Old Derby China Factory* (George Bell, London, 1876)

R. Jenkins, The silica brick and its inventor, William Weston Young. Trans. Newcomen Soc. **22**(1), 139–147 (1942)

E. Jenkins, William Weston Young, in *The Glamorgan Historian*, vol. 5, ed. by S. Williams (D. Brown Publishers, Cowbridge, 1968), pp. 61–101. ISBN 0-900807-8-1

E. Jenkins, *Swansea Porcelain* (D. Brown Publishers, Cowbridge, 1970)

S. Jenyns, *Later Chinese Porcelain* (Faber & Faber, London, 1971)

L. Jewitt, *Ceramic Art in Great Britain* (Virtue & Co., London, 1878)

W.D. John, *Nantgarw Porcelain* (Ceramic Book Co., Newport, 1948)

W.D. John, *Swansea Porcelain* (Ceramic Book Co., Newport, 1958)

W.D. John, *William Billingsley* (Ceramic Book Co., Newport, 1968)

W.D. John, G.J. Coombes, K. Coombes, *The Nantgarw Porcelain Album* (Ceramic Book Co., Newport, 1975)

A.E. Jones, Sir L. Joseph, *Swansea Porcelain: Shapes and Decoration* (D. Brown and Sons, Ltd., Cowbridge, 1988)

E.M. Nance, *The Pottery and Porcelain of Swansea and Nantgarw* (Batsford, London, 1942)

W.R.H. Ramsay, E.G. Ramsay, *The Evolution of Compositional Development of English Porcelains from the 16th C to Lund's Bristol c.1750 and Worcester c. 1752—The Golden Chain* (Invercargill Press, New Zealand, 2017). ISBN 978-0-473-409791

A. Renton, Thomas Pardoe and William Weston Young, in *Welsh Ceramics in Context,* vol. I, ed. by J. Gray (Royal Institution of South Wales, Swansea, 2005), pp. 120–146

S. Shaw, *The Chemistry of the Several Natural and Artificial Heterogeneous Compounds Used in Manufacturing Porcelain, Glass and Pottery* (Scott, Greenwood & Son, London, 1837). Re-issued in its original form in 1900 (1900), 713 pp

The Pottery Gazette, vol. IX, No. 92 (Organ of the China & Glass Trades, Stationers' Hall, Ludgate Hill, London, 1885)

W. Turner, *The Ceramics of Swansea and Nantgarw* (Bemrose & Sons, Old Bailey, London, 1897)

E. Young, *The Christian Experiences of Elizabeth Young, A Member of the Society of Friends, and Written by Herself*, ed. by W.W. Young (J.W. Wright & Co., Aldine Chambers, Bristol, 1842; and by Harvey & Darton, London, 1843)

W.W. Young, *The Diaries of William Weston Young, 1776–1847 (1802–1843)*, 30 vols., West Glamorgan Archive Service, Swansea, SA1 3SN. https://arcgiveshub.jisc.sc.uk/data/gb216-d/dxch/ddxch/i/hub

Chapter 2
William Weston Young: His Life

Abstract The development of William Weston Young's interests in ceramics with Lewis Dillwyn, decoration and manufacture of pottery and porcelain, his support of the efforts of Billingsley and Walker at the Nantgarw China Works, his association with Thomas Pardoe, other ventures, and finally his creation of the Dinas refractory brick.

Keywords William Weston Young · Elizabeth Young Dillwyn · Billingsley · Walker · Pardoe · Ceramics · Wrecks · Surveying · Salt extraction · Bankruptcy · Dinas refractory silica brick manufacturing

William Weston Young was born on the 20th April 1776, in Bristol, the third son of Edward Young, a corn factor in Lewins Mead, and Sarah, nee Weston, into a devout Quaker family. This was an auspicious year historically in that the Declaration of Independence of the thirteen American States from British rule was finally made on July 4th in the same year. He has been described variously as an artist, botanist, wreck-raiser, surveyor, potter and the inventor of the clay firebrick, as outlined above. He was sent to Gildersome School in Yorkshire, a Quaker boarding foundation which started in 1772 and closed down in 1815 with the death of its headmaster, at the age of 5 and he has been credited with the acquisition there of a rudimentary knowledge of science which eventually was put to good use in his invention of the fire-resistant refractory brick that was so essential for the lining of furnaces in the blossoming expansion of the British and world-wide iron and steel and copper smelting industries. In July of 1781, he sent his first letter home from his school, which he left at the age of 15 prematurely in 1791 because of the failure of his father's corn-milling business in Lewins Mead, Bristol. After two or three rather unproductive years as a shop assistant in Bristol, Young attempted to emigrate to the USA in 1794, at the age of eighteen, but his American-registered ship, the "*Severn*", was captured by the French off the south of Ireland and he was incarcerated, eventually he escaped from captivity and made an arduous journey home. He decided that he had enough of adventure by this time, so he settled in Bristol and married a fellow Quaker, Elizabeth Davies of Minehead, who was ten years his senior, on April 14th, 1795 at the age of 19, and set up a small local business venture with his brother, Edward. This, coincidentally, was the same year that William Billingsley made the decision to part company with

© Springer Nature Switzerland AG 2019

H. G. M. Edwards, *Porcelain to Silica Bricks*,

https://doi.org/10.1007/978-3-030-10573-0_2

William Duesbury II at the Derby China Works, where he had earlier been appointed the prestigious head of the decorating and enamelling workshop, and to set up the manufacture of porcelain at nearby Pinxton with John Coke. At this stage there is no evidence that Young and Billingsley knew of each other and, indeed, it seems that Young was now bent upon pursuing an agricultural trade as a miller, corn factor and farmer because in late 1797 he moved to Neath, in South Wales. In March 1800, he leased the Aberdulais watermill from John Llewellyn of Ynysygerwyn, Aberdulais, in the Vale of Neath, South Wales, for 21 years at a rental of £105 per annum; for this, he acquired financial support from his uncle Thomas Young. He lived in the picturesque Aberdulais watermill, described by him as "a beautiful property" and the subject of a later etching by Young in his Book "*A Guide to the Scenery and Beauties of the Vale of Neath*", published in 1835. A watercolour by the renowned artist J. M. W. Turner entitled "*Aberdylais Water Mill, Glamorgan*" is now in the Tate Gallery, London (acquisition number TW 1575; Wilton 169b; on loan to the National Library of Wales, Aberystwyth) and a print of the adjacent "*Cascade at Aberdillis, Vale of Neath*" by Henry Gastineau (ca. 1813) depicting the Dulais Falls over the Pennant sandstone outcrop at the confluence of the Dulais river with the River Neath is shown here in Fig. 2.1. He also leased the adjoining Forest Farm and the neighbouring Caer Pwlla for 14 years together at a rental of £47 5s. per annum. Young's chosen career in the agricultural business thus had a promising start, and he supplied a large catchment area in the Vale of Neath and the industrial complex of Merthyr, delivering corn on packhorses, but his large acquisition of corn and bean supplies in early 1802 at a cost of £1600 happened to coincide with the Treaty of Amiens, which marked the formal cessation of hostilities between Great Britain and the French Republic. This was signed on the 25th March 1802 (Germinal 4, Year X, in the French Revolutionary Calendar) and these particular commodities then suffered a sudden and significant depreciation in value. For example, Young had just purchased wheat for milling at 11s per bushel and overnight the selling price had dropped to just 3s per bushel. This, coupled with his innocent and perhaps rather naïve dealings with several suppliers who took advantage of the market unrest, resulted in Young being declared a bankrupt on May 8th, 1802, by a Commission in Bristol—the sum disclosed being some £6000, a significant portion of this being his wife's money. As a result, his leases on the Aberdulais Mill and on his farms at Caer Pwlla and Forest Farm were offered for sale. He and his family departed Aberdulais Mill with great personal sadness on January 20th, 1803. However, the Peace lasted only one year and on the 18th May, 1803, war broke out again between Great Britain and France. Young was very disconsolate that he and his family had received no support in his financial distress in the meantime from the Quakers in Neath, an active group which counted Joseph Tregelles Price as one of their founding fathers, and he promptly severed his connection with the Society of Friends, becoming a Wesleyan Methodist instead, although Elizabeth still remained a staunch Quaker for the rest of her life. Young's bankruptcy had a profound effect upon his career and he then made a momentous decision to have a career change.

On the 23rd January, 1803, William Weston Young and Elizabeth moved to lodgings in Swansea, only some eight miles or so from Aberdulais, where he took up

Fig. 2.1 Aberdulais Falls, where the River Dulais flows over a Pennant sandstone outcrop into the River Neath: print entitled *Cascade at Aberdillis, Vale of Neath*, drawn by Henry Gastineau for the multi-volume book, *Beauties of England and Wales*, dated 1813 and edited by J. P. Neale. The Aberdulais Mill House so beloved of William Weston Young and his wife is situated adjacent and to the right of the Dulais cascade. Private collection

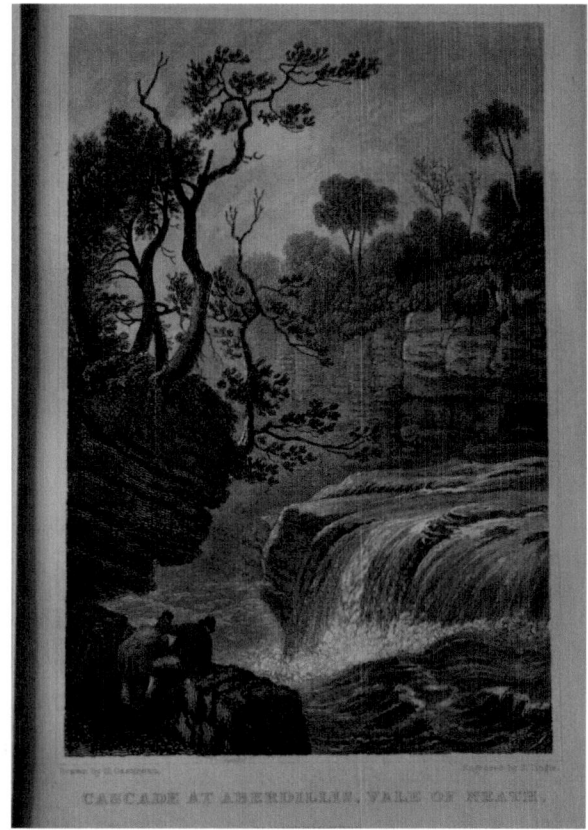

employment formally as a draughtsman on February 10th, 1803, with the young Lewis Weston Dillwyn, who was Young's junior by two years, at the former Cambrian Pottery, which had been re-named the Swansea Pottery, at a salary of £75 per annum. Both gentlemen had a common interest in natural history, although it is alleged that Young had previously decorated Swansea pottery whilst at Aberdulais as a side-line, and that he had installed there a muffle furnace for this purpose—if true, this surely started Young on his ceramics career at a very early age. Young, using his drawing skills, accurately depicted local flora and fauna species in detail with their taxonomic names for Lewis Dillwyn, who was researching these for a book; between 1802 and 1809 Dillwyn made a seminal study of British algae, which was published in 1809 as "*The British Confervae*", and Young's illustrations are to be found in Part III of this work, and thereafter. In Part IV, Dillwyn credits Young with the discovery of *Conferva dissiliens* (Plate 63) and *Conferva youngana* (Plate 102), for which Young was elected an Associate Member of the prestigious Linnaean Society, located in Burlington House, Piccadilly, London. It is indeed fortunate that William Weston Young kept detailed notes of his activities in diary form from 1802 until 1843 with several chronological absences, comprising some 30 volumes, which are now stored in the West Glamorgan Archive at Swansea (W. W. Young, *Diaries 1802–1843*).

Young stayed at Swansea for three years with Dillwyn until August 1806, from where he later moved to Newton Nottage, near Porthcawl, in October 1806. During this period at Swansea he met Thomas Pardoe, a ceramics decorator who was establishing a reputation for his quality of artistry and accurate depiction of natural history subjects, animals and botanical specimens on china. In the interim, Young's acumen for invention was further realised with his design for a grab-mechanism for the raising of wrecks (sometimes referred to as a *"forceps"* by contemporary observers); he rapidly put his invention to good use and by April, 1807, his salvage of a cargo of copper from the sloop *Anne and Teresa*, which had sunk in the Bristol Channel on a dangerous coast, established for him a change of fortune, rewarding him financially and essentially enabling him to discharge his earlier bankruptcy. Young achieved this by replacing a first-choice salvage expert, who had nevertheless failed to raise any copper ingots from the sunken ship. Elizabeth Young records in her Diary (Young 1842) that many attempts had hitherto been made to raise the cargo from the wreck, but all had failed. Young hired a number of workmen and a boat from Porthcawl and recovered an estimated two thirds of the copper cargo from the sunken vessel, for which he was paid a handsome £6000, representing a salvage cost of £45 per ton, of which his personal profit was said to be £1600. Elizabeth Young wrote in June 1807, that *"A year ago we were in low circumstances ... now, although not rich, yet we are made stewards once more of this world's goods"*.

Young then almost immediately commenced the recovery of the mixed cargo of the West Indiaman, *Trelawny*, which was wrecked on the treacherous Naze Sand in the Bristol Channel on December 11th, 1806.

His move to the village of Newton Nottage near the coast at Porthcawl, where he resided for 18 years from 1806 until 1824, prompted his starting to develop his new interest in land surveying, whilst still maintaining his interests in salvage work, farming, brickmaking and purveying general merchandise. Young's own diaries also record that he constructed in Nottage his own muffle furnace and ceramics kiln for the decoration and firing of pottery, which he bought in from the Swansea Pottery. The death of the local land surveyor, John Williams, in September, 1811, meant that Young was then poised to establish a base for himself there as a prominent surveyor in the county of Glamorgan and his location at Nottage was a prime location for the advancement of his new career, adjacent to the Margam Abbey estate of the Mansel Talbots, and close to the Nicholls of Merthyr Mawr, the Stradlings of St. Donats and the Ewenny estate of the Rice family, giving him access to local landowners and wealthy businessmen. However, he still retained links with Lewis Dillwyn after leaving Swansea and occasionally he performed decorating work for him at the Swansea Pottery; in February 1808, he also became a general merchant, selling Swansea earthenware, coal, bricks, tiles and agricultural produce from his home in Newton Nottage. A combination of his salvaging and surveying operations would later be seen as vital for his involvement in the securing of key business investors and landed gentry in support of the fledgling Nantgarw China Works. He quickly developed his land surveying business, with diverse contracts such as the surveying of coal pits and mines, an extension to the Neath Canal, surveying for boreholes and levels, and plans for the construction of a pier below Llantwit Major and a bridge at Clifton.

William Weston Young participated fully in the community life at Newton Nottage: he elected to go out to sea with local boatmen in six-oared boats to rescue the crews and passengers of stricken vessels sinking in the Bristol Channel. The later formation of the RNLI and organised efforts in lifesaving at sea only came in many years later, around 1860, with the establishment of the lifeboat *"Good Deliverance"* at Porthcawl, which put to sea as a 30-foot six-oared rowing boat with a crew of nine men, saving 28 lives from 17 rescues carried out between 1860 and 1872. It is rather reprehensible that Young's early humanitarian efforts in lifesaving have gained him the rather unkind appendages of *looter* and *wrecker* from some contemporary sources—perhaps, obliquely referring to his earlier exploits at salvaging cargoes from wrecks using his patented grab-machine! As referred to above, a major recovery operation undertaken in December 1806, by Young was the salvage of a mixed cargo of household items from the West Indiaman, *Trelawny*, which foundered on the Naze Sands in the Bristol Channel and the derogatory comments about his ulterior motives seem to have originated at around the same time, perhaps from dissatisfied, jealous and unsuccessful competitors. Elizabeth Young in her Diaries (Young 1843) states that he regularly put out in a boat with his crew to save lives at sea—amongst these she mentions that he rescued 12 sailors from the *"Trelawny"* from the ship's complement of 16, and others from an unnamed Portuguese ship with a cargo of fruit which foundered in the Bristol Channel. These life-saving efforts have largely been ignored by contemporary historians.

As a geologist, William Weston Young discovered the potential of a local limestone at Mumbles, Swansea, for use as a substitute marble for sculptures; in 1813, Thomas Rice Mansel Talbot died at Penrice Castle and Young was commissioned by his family to design and construct Talbot's tomb in Margam Abbey Church with the mandate of using only locally sourced minerals, which included Penrice alabaster and Mumbles marble. Clearly, Young was known to the family in his role as a land surveyor and that his artistic skills were also apparent to them. This tomb occupied much of Young's attention over the following six years and it was eventually completed in February, 1820, where it now stands in the nave of the Margam Abbey Church and a terminal invoice for over £168 for the design and materials exists in the Young archive held by his descendant, Colonel Young of Preswylfa, which was shown to Turner (*The Ceramics of Swansea and Nantgarw* 1897). Young's adherence to the completion of his commission for this tomb meant that for the initial setup and growth of the Phase II Nantgarw China Works, Young, who had achieved so much in gaining the interest and investment of local businessmen in the venture, was not able to participate fully in person at Nantgarw and his visits there to engage with William Billingsley and Samuel Walker were therefore mostly sporadic and quite irregular.

William Weston Young first met with Billingsley and Walker, according to his diary record, on January 30th, 1814, when they arrived at the small village of Nantgarw in the Taff Valley. Morton Nance has suggested that the catalyst for the bringing together of Billingsley, Walker and Young would have been Thomas Pardoe, who had known Billingsley at Derby and Young at Swansea; Young in his capacity as a land surveyor would probably have been acquainted with the advantages of Nant-garw as a potential site for the location of their project. At Nantgarw they proceeded to construct a factory for the manufacture of porcelain on the banks of the recently commissioned Glamorgan Canal (which had been commenced in August 1790 and reached Cardiff in 1794), built to transport iron, coal and goods some 25 miles from Merthyr (at that time the foremost commercial centre for heavy industry in Glamorgan) via the Rhondda Valleys at Pontypridd to the gateway deep-water port of Cardiff. The Glamorgan Canal was built by Richard Crawshay of Cyfarthfa and the Homfrays of Penydarren, the "Iron Kings" of South Wales, to transport their wrought iron from Dowlais, Merthyr to Cardiff to London. In 1813 some 100,000 tons of Dowlais iron was transported via the Glamorgan Canal as recounted in the *Directory of Cardiff, Caerphilly and Llandaff* by Thomas Ridd, who studied the use of the Glamorgan Canal in that year. No mention is made of the fledgling Nantgarw China Works in Ridd's survey, so it can be concluded that the China Works did not start up until 1814. This factory start-up enterprise is often referred to as the Phase I of the Nantgarw China Works. The location of the site at Nantgarw was critical for Billingsley and Walker since transportation problems in an emerging industrial Britain were acute: roads and turnpikes were often in a bad state of repair and the railway system which would revolutionise the carriage of raw materials and fin-ished goods between industries, ports and mines was still some two or three decades in the future—the Rainhill Trials in 1829, which was won by Robert and George Stephenson's "*Rocket*" steam-fired locomotive which achieved an average speed of 30 mph over five consecutive runs, established the British railway system and by the 1840s a network of railways carrying passengers and freight in successful commer-cial competition with the canals was operational. Billingsley and Walker arrived in Nantgarw after leaving their employment at the Barr, Flight and Barr Royal China Works at Worcester following the death of Martin Barr in 1813 on amicably good terms, although several authors have suggested otherwise, even implying that the selection of Nantgarw as the site for the manufacture of their new porcelain was cho-sen because of its isolation and the secrecy with which they wished to pursue their enterprise, and especially going into hiding from the perceived wrath of Martin Barr! The isolation of the Nantgarw site may well have been selected for other reasons, of course, and the choice of a quiet site far away from prying eyes and industrial espionage would have been an attractive proposition, but the financial parting gift of £200 from Martin Barr to Billingsley and Walker prior to their departure from the Royal Worcester China Works does not give credibility to the suggestion of their parting disagreeably with each other and it seems that their actual separation from Worcester was quite amicable. The reason for their departure will be investigated below, but Martin Barr had died late in 1813 and it seems that Messrs. Flight, Barr and Barr, taking over at the Royal Worcester China Works, were not keen to embark upon the formulation of new porcelain bodies. Martin Barr did insist, however, that Billingsley and Walker must not reveal the secret recipe and formulation of their new porcelain to a third party otherwise a legal penalty of £1000 would be sacrificed thereby—but there was no objection to them making porcelain of this secret recipe themselves! Some authors have made a point about the fact that Billingsley appeared to use the surname *"Beeley"* in the setup at Nantgarw to disguise his true identity. There is written documentary evidence for Martin Barr's stance towards Billingsley and Walker and for these statements, however, it is often alleged in earlier works that the real reason that Billingsley and Walker left the Worcester China Works in 1813 was the death of Martin Barr and the reluctance of his successors to engage in the production of new and experimental high-quality porcelains to the detriment of their existing product. As far as we know from historical documentation, William Billingsley kept his word and never revealed his successful recipe for the new Nantgarw porcelain to anyone, but it is equally clear that Samuel Walker did provide Taylor (*The Complete Practical Potter* 1847) with the secret recipe and formulation, which later even appeared in the *Cambrian* newspaper and the *Pottery Gazette* anonymously in 1885. In the late 1840s, Billingsley had been dead for almost twenty years and Walker was about to depart for a new life in the USA, where he is credited with setting up a ceramics factory at Temperance Hill, Troy, in New York state. There is now some evidence that Samuel Walker did continue with his ambition and that he did actually achieve the manufacture of porcelain in the USA at Troy, New York state (Broderick 1988), although it seems that the major output from this factory was functional brown-glazed earthenware of a *Rockingham* (more strictly, *Brameld*) type.

In the so-called set-up of Phase I Nantgarw, William Weston Young's financial dealings were rather complex and he is thought to have provided at least £600 of his own money to Billingsley and Walker in support of their initial enterprise, in addition to their own funds that were donated from Barr, Flight and Barr, at the Royal Worcester China Works. Young's free capital ran out in 1814, and he thought that it might be possible to obtain a grant from the British government to sustain porcelain manufacture at Nantgarw. Young persuaded Sir John Nicholl, an influential land-owner and Member of Parliament, to forward a *"Memorial,"* written by Billingsley, Walker, and himself, to the Secretary for the Lords of Council for Trades and Manufactures. This *Memorial*, dated September 5th, 1814, is reproduced here:

> To the Right Honourable The Lords of the Committee of Council for Trade.
>
> The late arrangement between this Country and France, in fixing a tariff on different Commodities and Articles of manufacture, and the probability of a commercial Treaty between the two Countries, has induced us, the undersigned memorialists to lay before your Lordships the following statements. It is now many years since France has taken the lead in the manufacture of porcelain - the Government of that country finding the benefit likely to result from it both as to immediate Profit to the Manufacturer, and also to the national reputation, has exerted itself to carry the Art to its highest Perfection, by encouraging able Men (including both Chemists and Painters) to turn their attention to the subject. English manufacturers have also exerted themselves in the Competition, and much Capital has been expended on Trials for the purpose of improvement; but that the success hereto has not been equal to the Exertion is sufficiently proved by the importation of white French Porcelain continuing to be a very considerable and increasing amount; the selling price of which for the last thirty years, has been near three times that of the best English white porcelain. This being the present state of the case, your Memorialists have to state to your Lordships that they have, either separately or unitedly have been engaged for some years (near twenty on the whole, but the last seven very closely) in trials for the improvement of British porcelain; and that they

have at length succeeded (with articles entirely of British produce) in making a porcelain equal in every respect to the French; and to the intent that your Lordships may be satisfied of the truth of their assertion they will, at any time when called upon, furnish a number of pieces both glazed and unglazed for your Lordships' inspection; or to be compared with any ware of French or any other manufacture. One or more of the Principals engaged in the manufactory will also be ready to attend your Lordships to explain anything necessary to be explained and give such further information as the case may require. Your memorialists have also to observe that the article they manufacture is not the result of fortuitous combinations, but is formed on true scientific principles, and they feel confident, if properly encouraged, that they will be able to make still further improvements;-also, with regard to the formation of their ware, they will undertake to make any article the French or any other people can make, and with as much taste and precision as they do; but, the manufacture being in an infant state, they are not yet furnished with those models that an older establishment must necessarily be in possession of. Your Lordships must be aware that it is often the case that men of real talent, capable of benefiting their fellow Citizens and the Nation at large, are so confined in their exertions for want of capital to pursue their object in an efficient way, that it sometimes happens, after a life spent in meritorious and persevering exertions, the benefit of their discoveries falls to the share of others; and not infrequently, the discovery itself dies with them, and is lost to the public. In these respects your memorialists have to state to your Lordships that, by great exertions of their own, and with the employment of a capital of about two thousand pounds, they have brought their manufacture into such a state as to produce about twenty-five dozens of ware (plates) per week; and they find their capital (so much having been expended in trials and erections) but barely adequate to this;-now this quantity, and, when compared with the quantity of white French porcelain imported, is very trifling; and, without some further assistance, they must be a long time before they can manufacture on such a scale as to do away with the necessity of importing French porcelain;-therefore, should the quality of their ware prove satisfactory, they submit to your Lordships the propriety of giving them encouragement and assistance in any way your Lordships may think proper, in order to forward the supply of the market with English porcelain. It is needless for your Memorialists to state to persons of your Lordships' judgment and knowledge, what great encouragement would also hereby be given to artists in a variety of the higher branches of painting and designing, as well as the tendency it would have to advance the national manufacturing reputation. Your memorialists having thus stated to your Lordships in as concise a manner as they can, the present state of the trade in general and of their manufactory in particular, pray your Lordships to take their case into consideration, and to assist them in any way it may appear to require. They beg leave to state to your Lordships, that on the conclusion of the Treaty of Amiens, the Government of France sent a service of porcelain as a present to Lord Hawkesbury, in order to exhibit their own superiority. - They would theretofore suggest to your Lordships the propriety of presenting a service of porcelain to some principal personage in France, in order to shew the advance made in the British manufacture. They also pray your Lordships to take into consideration the propriety of laying an additional duty on French porcelain, to act by degrees as a prohibition: -for that the present duty, tho' high, has not that effect is evident from the great importation of French ware, that which nothing can be a clearer proof of the state of the two manufactures.

Signed—Samuel Walker

William Beeley

William Weston Young

Manufacturers of Porcelain, Nantgarw-near Cardiff, Glamorganshire.

A detailed discussion of this *Memorial* and its implications is provided by Nance (*The Pottery and Porcelain of Swansea and Nantgarw* 1942). It is interesting that William Billingsley has signed this application for financial support in the name of

"*Beeley*" and a photograph of the handwritten signatures appended to the *Memorial* is shown in Fig. 2.2. Much has been read into the use of this diminutive by previous chroniclers and historians and the conclusion has invariably been that Billingsley was trying to disguise his presence in some way, perhaps escaping potential creditors and the influence of those who wished to determine and possibly undermine his porcelain manufacturing ideas and enterprise Personally, there seems to be no evidence for these conclusions at all, and Billingsley's movements were inexorably linked with those of Walker by this time anyway, so it would seem to be rather a pointless exercise to disguise Billingsley's name and not that of Walker simultaneously if secrecy were solely the desired objective. Certainly, when Lewis Dillwyn came to engage his services at Swansea some months later, his name had reverted to Billingsley again!

The Council, an advisory body to Parliament, met on September 12th to consider the financial request from the Nantgarw China Works; despite the eloquence of the "*Memorial*" submission in 1814 and for the rather modest sum of £500 of government funding requested for the new Nantgarw China Works, accompanied by specimens of the newly fired and decorated porcelain, Walker, Billingsley and Young were unsuccessful in their application. However, the attention of the President of the Council, Sir Joseph Banks, a well-respected scientist, adventurer, explorer and porcelain connoisseur was drawn to the inherent beauty and exceptional quality of the samples of Nantgarw china submitted therewith. He immediately alerted his friend and fellow ceramics enthusiast, whom he knew through the Royal Society, Lewis Weston Dillwyn, to the unsuccessful Nantgarw submission and suggested that he meet with Billingsley, Walker (and presumably, Young) to see if some collaborative venture for their mutual advantage could be forthcoming. Of course, Dillwyn and Young were already well-acquainted with each other from the years that Young had spent in the Swansea Pottery between 1803 and 1806 and through their mutual interest in local natural history. Several local meetings are recorded between Dillwyn,

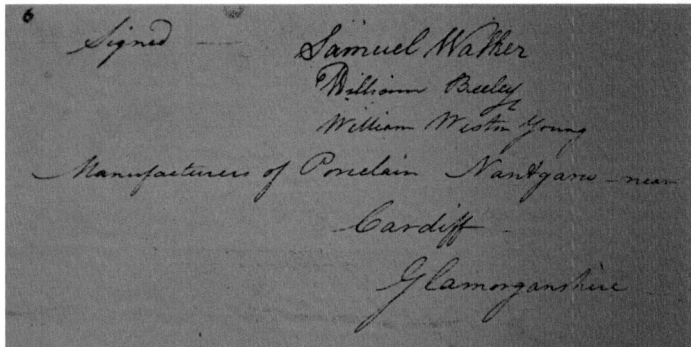

Fig. 2.2 Photograph of the signatories to the *Memorial* of September 5th, 1814, submitted in request for financial support for the emergent Nantgarw China Works by Samuel Walker, William Billingsley and William Weston Young. Note the diminutive appellation of Billingsley as Bealy, of which there has been much conjecture and debate by historians

Billingsley and Walker over the three weeks following the initial request of Banks to Dillwyn to establish contact and the result was that Billingsley and Walker were henceforth engaged later in 1814 to manufacture porcelain for Dillwyn at Swansea. It is intriguing that no one has recorded the presence of Young at these meetings, but his previous association with Dillwyn is suggestive that he could have had a major role in facilitating the arrangements. A statement within this *Memorial*, on lines 16 and 17, bears closer scrutiny as it refers to the applicants having been engaged in active experimental trials for the improvement of porcelain bodies for over twenty years hitherto and especially active in the previous seven years: this means that Billingsley and Walker, at least, had been trialling synthetic procedures for their china since 1807, which includes their time at Brampton-in-Torksey (Exley 1970)! This implies strongly that they were actually making porcelain at Torksey and not merely acquiring it from elsewhere for decoration, and wasters have been identified there of locally produced china of good quality and transparency (Chapman 2003)—and indeed, that their time at Worcester probably capitalised upon these experiments, refining their formulations and recipes for embarkation upon their final Nantgarw venture. Testament to the success of their porcelain venture at Brampton-in-Torksey is revealed by the discovery of a fine porcelain service in 2005, commissioned by a William Weston (1769–1833) an engineer of repute specialising in canal navigation who worked in the USA and in Great Britain—and who has been cited as one of the top 30 most inspirational engineers of the early 19th Century (Gardner 2010) and who has additionally been linked with Billingsley and Dillwyn.

The agreement between Lewis Dillwyn, William Billingsley and Samuel Walker at Swansea was formally signed on October 3rd, 1814 (Nance, *The Pottery and Porcelain of Swansea and Nantgarw* 1942, p. 251) and they physically transferred to Swansea from Nantgarw on October 8th. Young records this moment in his diaries as "Beeley and Walker left". It is interesting to speculate on the reason that William Weston Young was not involved in the engagement at Swansea—he may, of course, have been present when this was signed which makes his omission even more curious, given the support he had provided to the others at Nantgarw over the intervening months. It is certainly true that Young was in dire straits financially after the failure to secure support for the continuation of the Nantgarw China Works in the submission of the *Memorial*. He must have already been close to bankruptcy, as on December 17th, 1814, he was officially declared bankrupt for the second time. It is quite possible, therefore, that with this imminent change looming to his personal financial situation that his formal involvement with Lewis Dillwyn and the Swansea China Works would not have been a propitious undertaking for all concerned.

According to Henry Morris, an esteemed decorator at the Swansea China Works, in an interview with Colonel Grant-Francis on August 14th, 1850, which was later reported in full in the *Cambrian* newspaper on January 3rd, 1896, and thereafter by Turner (*The Ceramics of Swansea and Nantgarw* 1897), Dillwyn, who was in charge at the Swansea Pottery, had the desire to make porcelain at Swansea as early as 1811, when he engaged two former employees of the Coalport China Works, Messrs. Burn and Biggs, to achieve this objective. This was spectacularly unsuccessful, however, and the venture failed: it is revealed form Morris' statement to Turner that these new employees exaggerated their expertise and effectively were rather inept and totally unsuitable for employment in this regard. An interesting revelation in this interview with Henry Morris was that it was clear to him that Lewis Dillwyn considered Samuel Walker to be they principal in his dealings with the Billingsley/Walker ensemble regarding their proposed removal to the Swansea China Works—in fact, it was left to Walker to sign the agreement for their engagement with Dillwyn. Thereafter, there was much discussion between Dillwyn and Walker relating to improvements in the Swansea porcelain body and Billingsley did not feature at all in these discussions! The overall result of William Billingsley and Samuel Walker's removal to Swansea is that William Weston Young had again been left significantly financially disadvantaged by the failure of the Nantgarw China Works to gain official support from the British government, so much so that after his second bankruptcy in 1814 he reverted to pursuing his surveying business, the construction of the Talbot tomb at Margam, brickmaking and painting porcelain and earthenware utilising his private muffle furnace in Newton Nottage. At this point he is not on record as having any further contact formally with either Billingsley, Dillwyn or Walker at Swansea, and Young himself then became pre-occupied with the Mansel Talbot tomb design and its construction at Margam running alongside his land surveying duties. Yet, when Billingsley parted company with Dillwyn on December 23rd, 1816, to set up the so-called Phase II at Nantgarw, Young re-appeared on the scene, but he was still at that time officially bankrupt, a situation from which he was not discharged until 1818, yet he was still able to induce his business friends and associates in the county of Glamorgan to contribute funds to the revived Nantgarw porcelain manufacturing venture. In this context, a rather curious statement is found reproduced in earlier historical writings referring to the earliest manufacture of porcelain at Swansea after Dillwyn's engagement of Billingsley and Walker:

> The William Billingsley recipe was modified and improved at Swansea but this was still wasteful enough for Lewis Weston Dillwyn to abandon the project in Swansea and in 1817 the pair returned to Nantgarw.

Unfortunately, although seemingly stated with some plausible definition and source veracity, this does not meet with the facts, although it does seem warranted enough for serious historical consideration: for example, there is little evidence that Dillwyn used the Nantgarw paste recipe at all at Swansea in 1814, whether this had been provided by either Billingsley and/or Walker. However, a relevant statement exists in the literature that Billingsley and Walker were advanced a "loan" of £340 by Dillwyn upon their engagement at Swansea in 1814 to clear their debts at Nantgarw and for which they would construct two new kilns at Swansea. The basis for this statement is an account of the Chancery Court proceedings at the winding-up of the Swansea China Works in 1820 involving the Bevingtons. It is very likely, therefore, that some residual Nantgarw porcelain was decorated at Swansea in the earliest phase before these kilns became operational. It is generally believed that the first Swansea production was a "glassy" porcelain with a good translucency but a rather heavily potted precursor to the successful duck-egg porcelain which came later in 1816/1817. It is likely that these a few of these hypothetical Nantgarw remnants were decorated by Swansea artists such as William Pollard and their existence and identification has always been a matter for debate and conjecture.

Dillwyn's experiments on varying and improving the characteristic Swansea duck-egg paste body did not commence until mid-1815 and from an examination of his notebooks (Appendix B and Dillwyn 1922) it can be seen that his china compositional variations were certainly not based upon the Nantgarw formulation as the amount of bone ash used at Swansea was significantly less, Dillwyn's use of flint glass cullet was not pursued at all at Nantgarw, and several additives such as borax, smalt and arsenic oxide were unique to the Swansea recipes (Edwards, *Nantgarw and Swansea Porcelains: An Analytical Perspective* 2018).The statement also strongly implies that William Billingsley and Samuel Walker left Swansea together to start up again at Nantgarw—and this is definitely not the case, as Billingsley's departure from Swansea preceded that of Walker by several months—after Billingsley had departed Swansea in December, 1816, Walker remained behind until September, 1817, to assist Dillwyn in finalising his experimental variations on deriving a harder paste to replace his famed duck-egg porcelain, joining Billingsley in Nantgarw in September that year, shortly after Lavinia Billingsley had died in Nantgarw, and by then Walker's wife, Sarah, had died in Swansea in January, 1817.

It is interesting that for some nine months in 1817, Walker parted company with Billingsley as Walker remained at Swansea to help Dillwyn complete his experiments on the empirical variation in Swansea body formulations in an attempt to increase the robustness of his esteemed duck-egg paste. However, the result was the Swansea *"trident"* body, which contained a proportion of soaprock replacing that of china clay; this was deemed to be so markedly inferior in texture and translucency to the duck-egg paste variety that it was not acceptable to the discerning clientele who normally purchased Welsh porcelain from the London agents, such as Apsley Pellatt & Co. On September 28th, 1817, therefore, Dillwyn sold the Swansea Pottery and the Swansea China Works to Timothy and John Bevington, when thereafter they became known as Bevington and Co.: at the same time Walker returned to Nantgarw to join Billingsley and commence the re-generation of the Nantgarw Phase II operation there.

There have been several theories proposed as to the reason that William Billingsley departed Swansea in late 1816, leaving his daughter Sarah there, then married to Samuel Walker; a possible explanation is that Billingsley was still intent upon manufacturing the finest porcelain and Dillwyn had certainly come close to achieving this objective with his successful duck-egg paste—but Dillwyn's embarkation on its replacement by an ostensibly more robust but inferior soaprock body, which even if it had become commercially successful would have been seen by Billingsley as a dramatic reversal of this objective and a retrograde step in fine porcelain manufacture. Walker stayed on at Swansea as his kiln firing skills were necessary and vital for Dillwyn to achieve satisfactory ceramic firings from his experimental formulations—it is indeed possible that William Billingsley agreed to proceed ahead to Nantgarw to prepare the ground for the new start-up of Phase II there with Walker and Young, who now made a re-appearance on the scene. What is clear is that the arrival of Walker at Nantgarw some months later in September stimulated the production of porcelain there, the kilns were rapidly fired up and porcelain was made there shortly afterwards. In January 1817, Sarah Walker, William Billingsley's elder daughter, died in Swansea—by this time Billingsley had departed for Nantgarw with his younger daughter, Lavinia, and his housekeeper: a double tragedy struck later that year for William Billingsley when Lavinia died of a "cold" in Nantgarw in September 1817 (Edwards, *Swansea and Nantgarw Porcelain; A Scientific Reappraisal* 2017). In this book an interesting theory for the rapid demise of the two sisters, Sarah and Lavinia, was advanced in which their fatal symptoms were diagnostically attributable forensically to their previous employment as gilders, which would have exposed them to the toxic effects of mercury poisoning, giving rise to typical symptoms. Billingsley, describing Lavinia's illness and eventual demise in graphic detail to his wife Sarah, who was still resident in Derby, gives some clues as to the potential diagnosis of her problem and it certainly cannot be simply ascribed to a common cold.

For the start-up of Phase II at Nantgarw, William Weston Young had secured a consortium of investors, known as the "Ten Gentlemen of Glamorgan", who each invested £100 in the venture; the funds were collected by William Booth Guy of Dyffryn St. Nicholas as Young was still a bankrupt. This sum of £1000 was enhanced by a further £1100 contributed personally from William Weston Young, despite his status of being an undischarged bankrupt. This investment enabled Young, Billingsley and Walker to establish the factory at Nantgarw, to purchase raw materials and fuel for the kilns, to engage staff and provide a firm base for the distribution and sales of the desirable Nantgarw porcelain in the London market. It appears that Young did not become physically involved at the Nantgarw China Works until much later, leaving the production side to Walker and Billingsley. Up until July 1818, Young had been fully employed with the design and construction of the tomb for Thomas Rice Mansel Talbot at Margam Abbey, which lasted until March 1819. He last visited the Nantgarw China Works under the tenure of Billingsley and Walker on July 2nd, 1819. The last recorded Nantgarw porcelain decorated by William Billingsley was a set of six Masonic emblem mugs for a local purchaser in Cardiff, for which an invoice exists, dated December 1819. Relatively soon thereafter, and it is believed before or around April 1820, Billingsley and Walker left Nantgarw to be employed by John Rose at the Coalport China Works. On September 17th, 1820, Young met with Edward Edmunds, landlord of the Nantgarw China Works site, to arrange for the renewal of the lease on the China Works premises, which was agreed amicably.

In October 1820, William Weston Young then became the sole proprietor of the debt-ridden Nantgarw factory and he invited Thomas Pardoe, formerly of the Swansea Pottery and now established with his decorating business in Bristol, to come and decorate the existing remaining stock of Billingsley/Walker Nantgarw porcelain in the white for sale locally to recoup losses. On October 28th, 1820, Young inserted an advertisement for the first sale of Nantgarw china and assorted items in *The Cambrian* newspaper but from notes in his diary he had already decided to carry on with the manufacturing use of the site, hiring William Bedford for the building and maintenance of the muffle furnace and Thomas Jones, Thomas Pole and May Hewitt for the china preparation and decoration and David Jones for the grinding of the paste components and frit mixtures at his flour mill in Nantgarw. Billingsley had acquainted David Jones with the necessity and importance of an especially fine grinding of the paste components, especially of the calcined bone ash. Already, some china was being bought by noted local landowners such as the Crawshays, ironmasters of Cyfarthfa, Merthyr, and a fine dessert service had been commissioned by Wyndham Lewis of Greenmeadow, Llantrisant, an influential Member of Parliament, decorated in a very attractive rose pompadour ground colour with garden flowers by Thomas Pardoe. A very detailed account of William Weston Young's activities post-1820 has been given in Ernest Morton Nance's well-researched book, *The Pottery and Porcelain of Swansea and Nantgarw* (1942, Chap. XIII), where he recounts Young's tenure of the Nantgarw China Works between 1820 and 1822 and follows this with the subsequent history of the Nantgarw China Works site; this builds upon Young's published letter to the *Cambrian* newspaper on Friday, October 18th, 1822, in which he outlined the history of the Nantgarw China Works (Nance, *The Pottery and Porcelain of Swansea and Nantgarw* 1942).

2.1 The Three Nantgarw China Works Auction Sales

The First Nantgarw Auction Sale: this was announced in *The Cambrian* newspaper on Saturday, October 28th, 1820, and again on Saturday, November 4th, 1820. The sale would be conducted by Mr. John Aubrey at the Nantgarw china manufactory on Wednesday, the 8th November 1820.

> For Sale by Auction by Mr. John Aubrey
>
> At the China Manufactory at Nantgarw on Wednesday the 8th day of November, 1820.
>
> A Quantity of Nantgarw porcelain; a considerable portion of this enamelled and gilded in a superior style; it consists of several tea and breakfast sets, a number of plates, both dinner and dessert, some cups and a large assortment of ornaments of various patterns.
>
> A large assortment of white ware, glazed, both plain and embossed.
>
> A quantity of ware in the biscuit state; also, a variety of moulds fit for carrying on the business on a large scale.
>
> Tools and machinery used in the business.
>
> A quantity of saggars, bricks, Cornish Clay, sand, &c. Also, the household furniture consisting of beds, bedsteads, table and chairs, &c.
>
> This ware is well worthy of the attention of the Nobility and Gentry of the neighbourhood, as it may be the last opportunity of purchasing the Article made at Nantgarw. The other things may be viewed on the Monday and Tuesday, the 6th and 7th of November, preceding the sale.

This means that, since this sale was before the arrival of Thomas Pardoe at Nantgarw, this porcelain must have already been decorated locally by Billingsley and his co-workers, excepting of course the advertised white glazed and biscuit ware, and the glazed items would naturally have possessed the white No. 1 Nantgarw glaze. It is possible that Young himself also became involved in enamelling and decorating several items of the remaining stock, but this is perhaps implausible because of time restraints as he had been at Nantgarw for less than a month by this stage. It is more likely, therefore, that the decorated pieces offered in the First Auction Sale had already been enamelled by the task force at Nantgarw under Billingsley and Walker before they departed for Coalport—and this means that most of the items would also have been factory seconds or possibly defective in some way as the perfect porcelain would have been targeted for decoration and sale by their London agent, John Mortlock.

The Second Nantgarw Auction Sale: this was advertised in *The Cambrian* newspaper on the 28th April and the 5th May 1821—the sale to be held in Cowbridge on three days, namely the 9th, 10th and 11th May, 1821. This would coincide with the expectation of large crowds who would attend the annual auction sale of livestock there.

Nantgarw China for Sale

To be sold by auction by John Aubrey at his saleroom in Cowbridge on Wednesday the 9th day of May, 1821, and on the two following days.

A large assortment of Nantgarw China, beautifully enamelled, consisting of dinner and dessert plates, &c, &c, tea china, and ornamental pieces of various description; the whole amounting to several hundred lots. As the manufactory is not likely to be continued, this will be the last opportunity of purchasing this valuable china, which is universally acknowledged to excel in beauty all hitherto made. Likewise, will be sold a pair of beautiful furniture pictures painted on China.

Sale to commence each day at eleven o'clock in the forenoon, and the whole will be sold without the least reserve.

N.B. Specimens of the china may be viewed at the Auction Room.

No porcelain was offered for sale in the white, which probably means that Young had realised that it was not possible to achieve a realistic price for undecorated china (this comment had been made by William Billingsley when at the Pinxton China Works some years earlier); Morton Nance suggests that the china for sale was defective and would not therefore be attractive to purchasers in an undecorated state. There was no reference to gilded china in the sale and it seems that already economies were being made in the decoration at this stage. The china offered would have been predominantly decorated by Pardoe, and perhaps also by Young and other hands at the factory, such as May Hewitt. An interesting and relevant comment in the advertisement is: "*As the manufacture is not likely to be continued this will be the last opportunity of purchasing this valuable china, which is universally acknowledged to excel in beauty all hitherto made*".

The Third Nantgarw Auction Sale: announced in *The Cambrian* on the 19th and 26th October 1822, the third and final sale of Nantgarw porcelain:

Glamorganshire. To be sold by Auction by Mr. J. Aubrey (by order of the Trustees of Mr. W. Weston Young) on Monday, the 28th day of October, 1822, and on the following days until the whole be sold, on the premises at Nantgarw – a large assortment of Nantgarw porcelain richly painted and gilded; comprising dessert and tea services, ornamental china of various descriptions; also a quantity of the same porcelain in White consisting of dishes, plates, tureens, and a variety of other table ware; also Dessert and Tea ware, and a large Assortment of Ornaments.

It was intended that the sale of the above China should have taken place at Cardiff during the Race-week, but it having been impossible to complete the painting and gilding of several sets by that time, it has unavoidably been postponed to the period at present advertised.

The public are respectfully informed that the Proprietor of the Nantgarw Works having now given up the Manufactory, this will be the last opportunity afforded them of purchasing this beautiful Porcelain.

The Moulds, Saggars, and other Articles used in the Trade, will likewise be disposed of at the same time.

The sale to commence each day at eleven o'clock, and the whole to be sold without reserve.

Printed Catalogues may be had at the Cardiff Arms, Cardiff, and of the Auctioneer.

The wording of this advertisement is very revealing in that it indicates that a lot of china still remained in the white and undecorated and that Young had now it seemed relinquished any more thoughts about the manufacture of china at Nantgarw as the moulds and hardware were now being disposed of; this clearly implies that the disposal of this equipment could not have occurred at the First Auction Sale, as quoted frequently hitherto by Turner (*The Ceramics of Swansea and Nantgarw* 1897) and others (for example, Jewitt, *The Ceramic Art in Great Britain* 1878). It is to be noted, too, that the premises are not offered for leasing, yet Young's diaries actually do not record his paying rent for the China Works site into 1822. There are only two rent payments logged into Young's diaries for the period 1820–1822, and these are £15-15s on December 23rd, 1820, and £ 25-1s on May 3rd, 1821. Cardiff Race-week, originally targeted for the final sale, occurred from September 18th–22nd, 1822, so a considerable delay was incurred by Young and Pardoe in trying to complete the decoration of china sets for sale; apparently, the final painting was accomplished on the 18th October 1822, after which Thomas Pardoe left for Bristol. Pardoe then returned to Nantgarw on the 25th October, where he remained with Young until the Third Auction Sale, helping to pay off the workforce: Young then made over the stock remaining after the sale to Pardoe, who must have made an arrangement with the landlord Edward Edmunds to remain in Nantgarw House, where he continued to live until he died at Nantgarw on July 23rd, 1823. It is assumed that he decorated the remaining china in the white that remained at Nantgarw for subsequent sale locally to pay off the creditors. Effectively, therefore, Thomas Pardoe became the last proprietor of the Nantgarw China Works before its eventual closure later after his death in 1823. Morton Nance also suggests that the moulds and hardware remaining at the manufactory were still unsold in the final Third Auction Sale and that John Rose of Coalport visited Pardoe after Young had left Nantgarw to arrange the purchase of these items.

2.2   Thomas Pardoe at Nantgarw

It is evident that by this stage Young had given up on the idea of recreating the manufacture of Nantgarw china and all remaining stock must therefore have been Billingsley's production in origin. It also means that, as seen from his experiments, it is certain that Young had no knowledge of the Billingsley and Walker recipe and at this time there is a note to the effect that John Rose was in contact with Young, negotiating for the purchase of the moulds and other hardware, which Young obviously realised would no longer be useful to him. It appears that John Rose actually went to Nantgarw to purchase these items from Pardoe before he died in July 1823, but after Young had departed the scene. Thomas Pardoe joined Young at Nantgarw in January 1821, at first intermittently being based in his decorating studio at Bristol to conclude business there, but from February 20th, 1821, Thomas Pardoe moved to the Nantgarw China Works until he died there in July 1823, joined by his son, William Henry, who arrived there from Bristol on March 2nd, 1821. William Henry Pardoe also stayed at Nantgarw until after his father died and it is believed that William Henry Pardoe assisted his father in the decoration of porcelain at Nantgarw.

It is generally believed that it was not until February 20th, 1821, that Young finally met Thomas Pardoe for the first time since their time spent together in the enamelling workshop at the Swansea Pottery some 15 years earlier, whence Young departed in 1806, but in William Weston Young's Diaries, excerpts from which are given later in this book, it appears that Young and Pardoe did meet frequently in the intervening years. Pardoe was the man who would accomplish so much towards Young's endeavour to recoup the finances from the disastrous closure of the Nantgarw China Works factory in just the short time available to him. Their meeting in 1821, which occurred for the first time at the Nantgarw China Works site when Pardoe arrived there from Bristol, where he already had his profitable china decorating business, was so critical to the final operations there. It is conjectural to propose a reason why an apparently successful ceramics enamelling business operation being run by Pardoe in Bristol would be put at risk by his transfer to Nantgarw, and to wonder at the possible inducements which were offered by Young to entice him there, but perhaps Pardoe also recognised, as Billingsley had, the awesome beauty of the considerable porcelain stock remaining at Nantgarw for completion and decoration and the potential advantage this would confer upon his painting prowess and already high reputation as a ceramic artist. Pardoe's work is certainly highly appreciated by collectors and historians today, but it has not always been so, as Robert Drane, a well-respected local academic antiquary and historian, wrote a rather scathing attack upon Pardoe's "*mediocrity as a china painter*" as he perceived personally in William Turner's dedicated book on the Nantgarw and Swansea factories published in 1897 (*The Ceramics of Swansea and Nantgarw* 1897). It is fair to say that this view is definitely not held generally today and the contrary opinion now applies, that Young was rather fortunate indeed to secure Pardoe's services to decorate the remaining stock of Nantgarw porcelain for the auction sales in 1821 and 1822. We can but wonder what other inducement could have been offered to Pardoe by the impoverished Young, who had just purchased the stock and materials at Nantgarw in October 1820, as well as committing himself to leasing the factory site including all machinery from Edward Edmunds, the landlord at Penyrhos, which occurred on September 17th, 1820.

2.3   Porcelain Manufacture at Nantgarw, 1820–1822

Both Jewitt and Chaffers, early porcelain chroniclers, agree that the manufacture at Nantgarw ceased immediately after Billingsley and Walker left for Coalport in April, 1820 (Jewitt, *Ceramic Art in Great Britain* 1878; Chaffers, *Marks and Monograms on Pottery and Porcelain with Historical Notes on Each Manufactory* 1863) but they also raise the concern that the undecorated stock of glazed and biscuit porcelain surely would not have been great enough to occupy the full attention of an accomplished artist like Thomas Pardoe for two years! Turner (*The Ceramics of Swansea and Nantgarw* 1897) and Binns (*The First Century of English Porcelain* 1906) support this view but there is also a novel suggestion that perhaps Young did succeed in making a harder bodied porcelain which might have been analogous to that made by Dillwyn, namely his soaprock *trident* paste, at Swansea a year or two before. Williams (*An Investigation of the Nantgarw China Works Site* 1932) has hypothesised that Young had perhaps made a Billingsley and Walker formulaic paste which was more thickly potted but still recognisable as "Nantgarw"—but we now know from the inspection of Young's diaries that he had no knowledge whatsoever of Billingsley's paste composition, to the extent that he did not even realise that bone ash was a vital component of the Billingsley and Walker Nantgarw formula. We can now appreciate the observation made to William Turner by Richard Millward, who actually worked at the Nantgarw China Works site (Turner, *The Ceramics of Swansea and Nantgarw* 1897) relating to Billingsley's extremely close secrecy in mixing his paste components in his cellar, which was deemed to be strictly off-limits to other Nantgarw China Works personnel.

In this context, nevertheless, Turner states that a harder china must have been made at Nantgarw under Young and Pardoe's regime (*The Ceramics of Swansea and Nantgarw* 1897, pp. 68, 76) and even remarks, prophetically as it appears, that there might be in existence wasters of this hard paste bodied Nantgarw china buried somewhere on site! Frederick Litchfield (*Pottery and Porcelain* 1925) similarly alleges that Young's paste was harder and more vitreous than that of Billingsley—and we may well ask where these related comments originated if not in the recalled memory of someone who worked there post-Billingsley and Walker's tenure. It seems rather strange that several early chroniclers who relied on interviews with survivors from the work force at the Nantgarw China Works state that the manufacture of hard paste porcelain was carried out there at all, especially after Billingsley and Walker had departed. An alternative view is that Pardoe always painted on Billingsley's china, and none other, which then had a different glaze applied (the so-called Nantgarw No. 2 glaze devised by Young), which was creamy-white and more thinly applied than the Billingsley glaze. We may well ask how these conflicting statements arose—most interesting, forensically, is the fact that the earliest authors all allege strongly that a new Nantgarw paste had been created by Young, and it should be remembered that these same authors had access to personal statements made by the surviving ex-workforce at the factory. In a contrasting scenario, Herbert Eccles in 1914 pronounced that, from his examination of a considerable number of Nantgarw porcelain pieces submitted to the Special Loan Exhibition at the Glynn-Vivian Art Gallery in Swansea to celebrate 100 years of Welsh porcelain manufacture, that the Nantgarw porcelain composition and quality was "*invariant*" throughout its short history, unlike that of Swansea, even though he does assert that Young did carry out experiments at Nantgarw in January and February, 1821, in an attempt to emulate the *trident* body recipe which had been the downfall of the Swansea China Works! This statement is borne out by a reference to Young's Nantgarw body recipe in Turner (*The Ceramics of Swansea and Nantgarw* 1897, p. 326) and to a recipe there for a soaprock bodied porcelain with a frit composition comprising 4 parts sand, 1 part soaprock, ½ part alkali, to 5 parts of which 1 part of china clay was added! It is highly significant that this recipe for Young's so-called Nantgarw porcelain completely ignores the presence of bone ash which was so critical in the Billingsley and Walker formulation—again supporting the idea that Young had no knowledge of the presence of this vital component. Several statements from ex-employees at the Nantgarw China Works allude to a secretive William Billingsley having a cellar which he used personally to effect the mixing of his recipe components, to which no one else had access as stated above, and this would confirm the supposition that even a valued colleague and partner such as William Weston Young may not have been privy to the finer details of the formulaic recipe for the original Nantgarw porcelain paste. Yet, in contrast, it appears that Samuel Walker must have had access to Billingsley's formulation, as he revealed this to Taylor (*The Complete Practical Potter* 1847) some years later before emigrating to the USA.

Morton Nance raises a very provocative question when he suggests that in his opinion William Weston Young purchased Swansea soaprock porcelain (known alternatively as the "*trident*" paste version) which he then decorated and glazed at home in his muffle furnace and sold as Nantgarw—which, if sourced from the factory locally would certainly have been assumed by clients to have been manufactured there. The really interesting corollary to these many and conflicting statements and opinions, of course, is that they could all relate back to a distant source memory of the production of a soaprock or harder variety of Nantgarw porcelain at Nantgarw, which would be tantamount to its being classified as "hard paste" because it would be devoid of a phosphorus component provided by the addition of bone ash. Then, we can alternatively point to the contrary opinion of Herbert Eccles who expressed in 1914 that Nantgarw porcelain was of a "constant, unchanging composition"—which may well have been be true for the Nantgarw specimens submitted to the Glynn-Vivian Art Gallery Loan Exhibition (GVLE 1914) on the basis of which Eccles made his statement, but forensically that does not mean that there are other specimens of Young's variant out there somewhere which could easily have been rejected visually as Nantgarw hitherto because of their hard paste composition?

Whatever can be gleaned from these contradictory and frankly confusing allegations, it has been unequivocally stated by most modern authors that no porcelain was made at Nantgarw after 1820, in fact accepting the opinion expressed by Herbert Eccles in 1914 as evidence of this; as will be discussed below, this statement is open to misinterpretation and it should not be inferred that attempts to manufacture porcelain were not undertaken there, however unsuccessful commercially these may have been. It should be noted here that Eccles never actually analysed any Nantgarw porcelain from the Special Loan Exhibition and his only analytical work at that time on the porcelain items in this Exhibition was undertaken on a single Swansea porcelain plate from the *Gibbins* service, which itself raises some difficulties now because of its belonging in a timeline to the T & J Bevington tenure period at Swansea, after Dillwyn had left the factory, and the Bevingtons too were engaged in a clearance sale of remaining stocks of porcelain, including perhaps some experimental examples from Dillwyn's own attempts to synthesise a more robust porcelain body. Eccles' analysis conclusively demonstrated that this particular Swansea plate does not conform in composition with that of specimens of conventional Swansea duck-egg porcelain that were analysed later.

In fact, the diaries and operational workbooks of William Weston Young clearly demonstrate that he did attempt to recreate Nantgarw porcelain there in 1820 and 1821, and he commenced this undertaking in October, 1819—these were considered unsuccessful as he did not have Billingsley and Walker's Nantgarw porcelain recipe, although he must have tried to guess their composition. Nance (*The Pottery and Porcelain of Swansea and Nantgarw* 1942) is rather ambivalent on the one hand in his assertion that William Weston Young made porcelain at Nantgarw and, alternatively, that it was highly improbable that he did do so. He states: "*We shall see later that some writers have maintained that during Young's tenure of the Nantgarw factory he himself made a paste which was much harder than Billingsley's Nantgarw body and resembles the trident paste of Swansea. It is however most improbable that Young succeeded in making any body at all at Nantgarw, At the most his efforts were confined to an endeavour to reproduce Billingsley's original body which he had always admired and which he knew to be saleable if correctly turned out. Certainly, he would not have attempted to reproduce a body the same as trident for in 1821 that had already proved to be a failure in the London market. Isaac Williams is very clear in asserting that William Weston Young did make porcelain at Nantgarw*". As will be seen later several of these apparently contradictory statements actually do bear some weight from what we know Young did during his tenure at the Nantgarw China Works.

What is evident and well-documented is that Young did perfect a new glaze for use in the decoration of the Nantgarw porcelain remaining in stock at the factory: the so-called Nantgarw No. 2 glaze is now referred to as the "*Pardoe glaze*" and is much creamier than Billingsley's original, with a sometimes more pitted, irregularly applied appearance. Thomas Pardoe died in July, 1823, and presumably the remaining stock of already decorated porcelain was sold at auction in 1823–1824 along with the remaining Swansea stock from the Swansea China Works. It is intriguing that Turner (*The Ceramics of Swansea and Nantgarw*, 1897) states that William Weston Young actually started experiments in creating Nantgarw china in January and February, 1821, but these were judged to be unsuccessful for the body but much better for a glaze. Young records in his diaries that on February 24th, 1821 he fired up "*our own glaze and it was good*" and his last recorded experiments on March 31st, 1821, brought the comment that it was "*the best glaze*". It appears, therefore, that there were several continuous but rather unsuccessful efforts to restart Nantgarw china production initiated by William Weston Young.

Binns (*The First Century of English Porcelain* 1906, p. 219) claims to have identified five distinct and definitive periods of Nantgarw porcelain production, namely:

China made by Billingsley before leaving for Swansea in November 1814 (Author: this date is incorrect as it is recorded by Young that Billingsley and Walker left Nantgarw for Swansea on October 8th, 1814).

Nantgarw china made at Swansea.

Nantgarw china made at the Nantgarw China works on the return of Billingsley and Walker to Nantgarw in 1817.

William Weston Young china made at Nantgarw: disputed by Morton Nance but supported by Isaac Williams.

Nantgarw china made at the Coalport China Works and *marked* Nantgarw: supported by Morton Nance who has identified this type with the presence of a red enamelled *Nantgarw* script mark.

Although several points here are debatable but we now should add to this list a sixth period, namely, china made at Nantgarw by William Henry Pardoe in 1857–1858. It seems definite, therefore, that the historical concept that Nantgarw china was made only at Nantgarw during the years 1817–1819 is simplistic and that possibly modern analytical science can illuminate the conjectural hypotheses and establish credibility for several conflicting statements.

2.4   The Birth of the Silica Refractory Brick

Meanwhile, Young had reverted again to his inventive mode upon immediately leaving the Nantgarw China Works. Firstly, he developed a smoke sensor for the smelting of copper at Swansea, as a result of a movement for the abatement of the foul air effluent which was deemed necessary by the local council for the good health of visitors and the local populace, but he did not get awarded the funding prize of £1000 by the assessment committee to proceed, although he did make the final selection of four potential remedies on the short list. The vetting committee decided that a suitable project had not been submitted and as a result no prize was awarded.

Young then had the idea of making superior firebricks for blast furnaces using a raw material with a high silica component to increase their hardness and durability. His first recorded experiments in firebrick trials were actually undertaken in the Nantgarw kilns and his formulation and firing recipe was finalised in 1821. In February, 1822, Young leased land from the Marquis of Bute for a period of 21 years near Pontneddfechan (transation from the Welsh; *Bridge over the Little Neath*), at Craig-y-Ddinas (translation from the Welsh: *Fortress Rock*) in the Vale of Neath where he had found a source of highly superior fireclay, which analysed as virtually pure silica (>98%). He had by now left his home in Newton Nottage and was based in Fairyland House, Neath, to be much closer to the site of his new business venture. Although he had acquired the lease and patent No 5047 for his silica firebrick recipe he still had to seek the funds to develop the site of the new Dinas Firebrick Company, which eventually was established at Pont Walby, about one mile down-river from his primary fireclay geological source at Craig-y-Ddinas, on October 19th, 1822. The Marquess of Bute visited the property with Young on October 17th, 18th and 19th, 1822, to conclude the terms of Young's lease of the property. There are reports that a brickmaking site already existed at Pontneddfechan from the 1780s, and possibly this was the one associated with Messrs. Fredericks and Jenner, but it is not clear if this constituted the same site developed by Young for his own enterprise there. Young's co-proprietors in the Dinas Firebrick Company were David Morgan (a Neath ironmonger), John Player and his elder brother, Joseph Young—the company was very successful and sold firebricks all over the world. A document exists, dated February 2nd, 1822, and signed by David Morgan, ironmonger of Neath, and William WestonYoung of Newton Nottage, land surveyor, which reads as follows:

> That he (Morgan) shall not make bricks or any other article from the same sand (Dinas) without the consent of William Weston Young, who does also agree on his part to bind himself in the same manner and under the same penalties (£5000) not to reveal the said method to any other person or persons or to make any firebrick or other article for sale from the said sand without the consent of Mr. David Morgan but that he shall be at all times at liberty to use the said materials for his own purposes at his china manufactory or elsewhere.

This is a very revealing document as it implies that even as late as 1822 William Weston Young still had it in mind to make porcelain—at Nantgarw or elsewhere? The Dinas Firebrick Co. eventually passed via Joseph Young to his son, William Weston Young Jr. (1798–1866), the name of the Company having morphed meanwhile into John Player and Co. in 1825, Riddles, Young and Co. in 1829, and finally Young and Allen in 1852. The Young in these latter companies was the eponymous nephew of our William Weston Young. When a metallurgist, John Percy, visited the brickworks to research silica bricks for furnace applications in 1859 he met an Edward Young, who was then the proprietor, but he does not mention a William Weston Young Jr. The brickworks closed in 1964 under the ownership of Richard Thomas & Baldwin, after mining of the Dinas rock had ceased in 1921; it was claimed that firebricks were supplied to the Swansea White Rock Copper Company for over 40 years for their smelting of copper ores locally. The fortunes of William Weston Young's nephew, called William Weston Young the Younger, took a downturn just a few years after Riddles, Young and Co. took over the brickworks in 1829 and by 1834 they went into receivership with the breakup of the partnership as noted in the following extract from the *London Gazette* of January 1834:

> Notice is hereby given that the Partnership heretofore subsisting between us the undersigned Thomas Hooper Riddle, Herbert Riddle, John Matthew Young and William Weston Young the Younger, in the business of Manufacturers of the Dinas Firebrick carried on by us at Neath, in the County of Glamorgan, under the style or firm of Riddles, Young and Co., is dissolved and determined as regards the said William Weston Young who retires therefrom, all debts due and owing to or from the late Copartnership are to be received and paid by the continuing Parties, by whom the said manufactory will be carried on. Dated this 11th Day of January, 1834.

> T. H. Riddle, Herbert Riddle, J. M. Young, William Weston Young, Jr.

This entry is preceded by a similarly worded entry notification of the dissolution of the same Partnership for the business of Wholesale and Retail Provision Dealers carried on in Neath, also under the description of Riddles, Young and Co., but here the intention of William Weston Young the Younger to carry on trade as a sole person in Wholesale and Retail Provisions is noted. Clearly, therefore, William Weston Young Jr. was now intent upon relinquishing all interest in the Dinas silica brickworks and instead devoting his attention to the general wholesale business in Neath. It seems likely therefore, that Edward Young had stepped in to replace William Weston Young Jr. and had taken over the business by the time that Percy visited the company in 1859.

In 1897, Turner (*The Ceramics of Swansea and Nantgarw* 1897) noted that "*bricks of Dinas sand are still being made by the Vale of Neath Firebrick and Cement Co. Ltd.*", of which the managing director was Colonel Young of Preswylfa, Neath, the great-nephew of William Weston Young. Turner also comments that: "*Silica bricks invented in the Vale of Neath are being used in the Siemens and Bessemer converters. The celebrated Dinas firebricks stand unequalled for application to intense and long-continued heat. The extraordinary and exceptional fireproof qualities of these bricks have secured them the first position in England, France, Germany, Russia, and in North and South America. Placed in furnaces above the wash of the molten metals and above the clinker line in the grates they display their qualities to great advantage*". He concludes his statement with an assessment of Young as "*The inventor certainly did a good thing for our manufactories even if he did not benefit much himself personally form their production*". Turner's remarks are particularly appropriate for the Dinas silica firebrick production even if some 40 years later, Nance (*The Pottery and Porcelain of Swansea and Nantgarw* 1942, p. 152) sums it up in just a single line of text as "*William Weston Young's invention of the firebrick was successful in contrast to his salt extraction from the brinefields of Stoke Prior, Worcestershire, which was very unsuccessful*".

2.5   William Weston Young's Interests Post-Silica Brick Invention

Even whilst building up the successful silica brick manufacturing business, William Weston Young Sr. still pursued his other interests, as exemplified by the granting of a patent, No. 5047, granted on the 4th December 1824 (*Newton's London Journal of Arts and Sciences* 1824):

> Patent awarded to William Weston Young of Newton Nottage, engineer, for his discovery of certain improvements in the manufacture of salt, part of which improvements are applicable to other useful purposes.

The basis of his invention was a brine evaporating pan which was placed over a fireplace and sited underneath an iron dome connected by a flue with the space below a second pan supported on iron bars and bearers: the steam from the lower pan is used to evaporate the contents of the upper pan. Young put this patent to use in establishing a salt works at Droitwich in Worcestershire. Jenkins (1942) thought it curious that Young's invention for salt extraction could not be correlated with any South Wales industry and suggested that Young's contacts in Bristol had engaged with salt extractors in Worcestershire, alerting Young to the need for his expertise in this venture. Young established himself at Stoke Prior, outside Droitwich, and drilled for salt deposits there—which according to Turner (*The Ceramics of Swansea and Nantgarw* 1897) he missed by a only a few yards: it is now known that geologically the salt deposits at Stoke Prior are much deeper than the corresponding strata at Droitwich and it seems that Young had not sunk his shaft deep enough and was unsuccessful in reaching the salt. In 1829, he returned to Neath where his interests are unknown: his wife in her Diary (Young 1842) hints at a business venture with "an opening has presented itself at Neath" and that this was possibly connected with the Dinas Brickworks? This business took him away at times over the next few years, as Elizabeth Young comments in her diary for March 1834, that "*he became unwell with suspected cholera and he was much changed as a person*" (Young 1843). Although still resident in Neath, they had by now moved into the centre of town from Fairyland House for convenience and possibly for Elizabeth to be closer to her Friends Meeting House and there are reports in his diaries that he was engaged in purveying general merchandise there. It was around this time that William Weston Young persuaded friends and business associates to invest several thousands of pounds in developing the Gnoll Colliery at Neath, which unfortunately was a unsuccessful, as was another of his ventures, namely, the design of a horizontal driving windmill which could be an alternative power source to steam power or water power for mill engines.

In his later years, William Weston Young settled in Bristol from 1835 and it is clear that he had still not given up on his dream of making Nantgarw porcelain there, presumably with smaller kilns that he had constructed for that purpose. In the same year, John Wright of Bristol published Young's book "*A Guide to the Scenery and Beauties of Glyn Neath*" (Young 1835), with illustrations by Young – the cover page of which is shown in Fig. 2.3—and Young personally supervised its production. In the Preface to this book, Young writes "*The drawing, etching, colouring and writing are all done by my own hand*". A version of this same book appears in the same year from the publishing house of Longmans, Rees, Orme Brown and Green of Paternoster Row, London, in 1835. An appreciation of Young's work has appeared in a recent book, entitled *An Attempt to Depict the Vale of Neath in South Wales: An Artistic Impression* (Yerburgh 2001), illustrating a pictorial journey around the Vale of Neath by the author as undertaken by William Weston Young in 1835.

Although Young's porcelain project seemed to be in abeyance, it was not forgotten and Young was still convinced that he could deliver the idea of a recreated Nantgarw porcelain. In partnership with John Wright and with operations conducted at Wright's establishment at Bristol, it seems that quantities of a porcelain called "*Nantgarw dry mix*" were marketed in 1837, to which the addition of water would make a porcelain slip paste. Young decided to tour the Staffordshire potteries and attempts were made to generate sales with factories in Stoke-on-Trent such as Spode, and also a start-up factory in London was mentioned, but the venture did not succeed commercially. We

Fig. 2.3 Cover page from William Weston Young's "*A Guide to the Scenery and Beauties of Glyn Neath*", published in 1835 by Longmans et al., Paternoster Row, London, with illustrations of local scenes by William Weston Young. The particular scene illustrated here shows the ruined 12th Century Norman Neath Castle, which still stands in the centre of the town, close to Young's residence in "Fairyland" when he was involved with the manufacture of Dinas silica bricks in the Vale of Neath

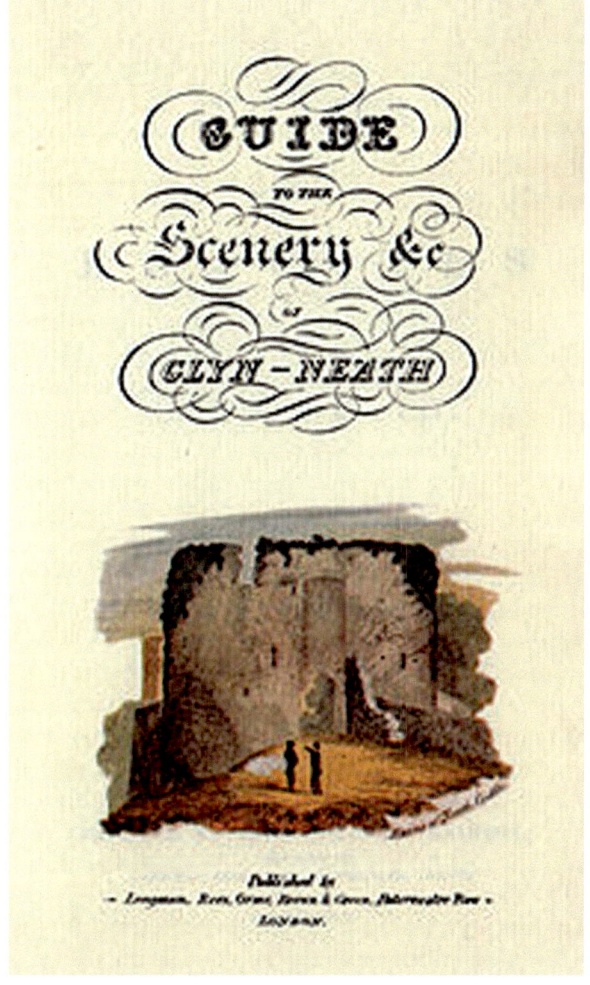

do not have precise knowledge of Young and Wright's composition of this "*Nantgarw dry mix*" but correspondence exists which indicates that several porcelain factories who purchased this item speculatively were not impressed with the material that they received: it is reasonable to suppose that potential buyers of this "*Nantgarw dry mix*" were naturally expecting it to give a porcelain body after admixture with water and subsequent firing that was similar to the original Nantgarw porcelain, and in that case they would have been severely disappointed if the experimental hard paste variety of Young derived from his experiments in 1820 and 1821 was offered for sale in this way. One thing is clear, from the timing of this venture it is certainly not correct to even consider that Young and Wright would have suddenly discovered the secret Nantgarw recipe of Billingsley and Walker which they were then intent of marketing themselves. Family fortunes took a turn for the worse in 1840 when Elizabeth Young became bedridden for 15 months as a result of a hip fracture sustained from a fall, and she died two years later in 1842, when Young was 66 years old. She is buried in the Society of Friends cemetery in Redcliff, Bristol.

When William Weston Young died of apoplexy on March 5th, 1847, aged 71, he was in post as the colliery manager of Alvely Wood Coal Works in Kidderminster; he is buried in Lower Mitton, Stourport. It is intriguing to suggest, without any evidence for or against, that a possible cause of the fatal apoplectic fit suffered by Young in 1847 could have been the imminent revelation the very same year by John Taylor of the true Nantgarw paste formulation ascribed to Samuel Walker which, of course, was totally different to that which had been tried by Young and Pardoe at Nantgarw some twenty-five years earlier—especially the mention therein by Taylor of a highly significant bone ash component which had not featured at all in Young's trials at Nantgarw. It could be argued that the revelation of this vital component which had completely escaped the attention of Young and had eluded him for some twenty-five years would have certainly caused him some considerable distress and could well have contributed to his demise had he been made aware of it. The question remains as to why Billingsley and Walker had never apprised Young of their formulaic recipe for the Nantgarw soft paste porcelain: he was an equal business partner and had done so much to secure the necessary funding for the setting up of Phase II of the Nantgarw China Works that it seems churlish to exclude him from this vital information, especially since when Billingsley and Walker left Nantgarw in 1820 they did so with a clean slate—Young having agreed to take on all the debts associated with the enterprise, becoming bankrupt again eventually in the process. A rather obvious reason for Billingsley and Walker not informing Young of the Nantgarw porcelain recipe and formulation is that thereby Young was protected from any threat of legal intervention being called by Flight, Barr and Barr of the Royal Worcester China Works against him or them for revelation of the details of the composition of the manufacture—although for all their time resident at Nantgarw between 1817 and 1820, there was never any suggestion of this happening, presumably because they were manufacturing the porcelain themselves. By transmitting the recipe to Young after leaving Nantgarw they would have immediately been liable to prosecution if Flight, Barr & Barr were intent on promoting this action and it could be argued that Young would have suffered this fate too especially if he then proceeded to re-generate the production of Nantgarw porcelain during his tenure there between 1820 and 1822.

The only known image of William Weston Young is a silhouette made of him of indeterminate age which is in accord with his Quaker principles, even though his perceived disappointment at the lack of support received from local branch of the Society of Friends for him in his darkest hours did seem to alienate his beliefs. He has been described (Jenkins 1942; Nance 1942) in his early 40s as a tall, fair and fine-looking man and possessed of great physical strength. He was widely read and encompassed an artistic as well as scientific and engineering expertise, especially in geology, mining and invention. He did not acquire wealth: the routines of manufacturing business did not appeal to him and he preferred to engage with the solution of technical problems, which after solution he would leave to others to develop. A classic example of this is his remarkable success in the invention of the silica refractory brick, which having seen through to a commercially viable entity was then lost to his interest and he left to others to its manufacture and to reap the rewards.

References

W.M. Binns, *The First Century of English Porcelain* (Hurst & Blackett, London, 1906)

W.F. Broderick, An English porcelain maker in West Troy. Hudson Valley Reg. Rev. **5**(2), 23–40 (1988)

W.B. Chaffers, *Marks and Monograms on Pottery and Porcelain with Historical Notes on Each Manufactory*, 1863, *J. Davy & Sons, London*, Kessinger Legacy Reprints (Kessinger Publishing, Whitefish, 2010)

R.E. Chapman, William Billingsley at Brampton-in-Torksey, in *Welsh Ceramics in Context*, vol. I, ed. by J. Gray (Royal Institution of South Wales, Swansea, Gomer Press, Llandyssul, 2003), pp. 179–192

L.W. Dillwyn, *British Conferva or Coloured Figures and Description of British Plants Referred by Botanists to the Genus Conferva* (W. Phillips. George Yard, Lombard St., London, 1809)

L.W. Dillwyn, *Notes on the Experimental Production of Swansea Porcelain Bodies and Glazes*. Made by Lewis Weston Dillwyn with Samuel Walker at the Swansea China Works Between 1815 and 1817. Presented to the Library of the Victoria &Albert Museum, South Kensington, London by John Campbell in 1920. Reproduced in Eccles & Rackham, *Analysed Specimens of English Porcelain*, 1922, see reference below

H. Eccles, B. Rackham, *Analysed Specimens of English Porcelain in the Victoria and Albert Museum* (Victoria and Albert Museum, London, 1922)

H.G.M. Edwards, *Swansea and Nantgarw Porcelain: A Scientific Reappraisal* (Springer, Dordrecht, 2017)

H.G.M. Edwards, *Nantgarw and Swansea Porcelains: An Analytical Perspective* (Springer, Dordrecht, 2018)

C.L. Exley, *A History of the Torksey and Mansfield China Factories* (Keyworth & Fry Ltd., Lincoln, 1970)

P.T. Gardner, *Billingsley, Brampton and Beyond—In Search of the Weston Connection: The Provenance of a Porcelain Service over 200 Years Old is Investigated* (Troubadour/Matador Publishers, Leicester, 2010)

R. Jenkins, The silica brick and its inventor, William Weston Young. Trans. Newcomen Soc. **22**(1), 139–147 (1942)

L. Jewitt, *Ceramic Art in Great Britain* (Virtue & Co., London, 1878)

F. Litchfield, *Pottery and Porcelain A Guide to Collectors*, 2nd edn. (A. & C. Black, London, 1925)

London Gazette, Part I, Friday January 17th, 1834, No. 19120 (1834), p. 101

E.M. Nance, *The Pottery and Porcelain of Swansea and Nantgarw* (Batsford, London, 1942). *Newton's London Journal of Arts and Sciences*, Patent Issue, p. 558, 4th December 1824

J. Taylor, *The Complete Practical Potter* (Shelton, Stoke-upon-Trent, 1847)

The Pottery Gazette, vol. IX, No. 92 (Organ of the China & Glass Trades, Stationers' Hall, Ludgate Hill, London, 1885)

W. Turner, *The Ceramics of Swansea and Nantgarw* (Bemrose & Sons, Old Bailey, London, 1897)

I.J. Williams, *The Nantgarw Pottery and its Products: An Examination of the Site* (The National Museum of Wales and the Press Board of the University of Wales, Cardiff, 1932)

D. Yerburgh, *An Attempt to Depict the Vale of Neath in South Wales: An Artistic Impression* (D. Yerburgh, Salisbury, 2001)

W.W. Young, *The Guide to the Scenery and Beauties of Glyn Neath* (John Wright & Co., Bristol, 1835) (sold by Longman, Rees, Orme, Brown & Green, Paternoster Row, London)

E. Young, *The Christian Experiences of Elizabeth Young, A Member of the Society of Friends, and Written by Herself*, ed. by W.W. Young (J.W. Wright & Co., Aldine Chambers, Bristol, 1842; and by Harvey & Darton, London, 1843)

W.W. Young, *The Diaries of William Weston Young, 1776–1847 (1802–1843)*, 30 vols., West Glamorgan Archive Service, Swansea, SA1 3SN. https://arcgiveshub.jisc.sc.uk/data/gb216-d/dxch/ddxch/i/hub

Chapter 3
The William Weston Young Diaries

Abstract A description of the written records of William Weston Young's activities as revealed in his Diaries, which although incomplete give much information for research into the chronology of events and his association with key partners in his business.

Keywords William Weston Young diaries · Insight into chronologies · Nantgarw China Works · Dinas brick manufacture · Wreck raising · Salt extraction · Porcelain manufacture

During this historical survey it is apparent that William Weston Young's life was very closely linked with several other people, notably William Billingsley, Lewis Dillwyn, Thomas Pardoe and Samuel Walker, whose interaction with him, both singly and in combination, was critical towards the successful outcome of many of his scientific and business activities. It is appropriate, therefore, that we now consider briefly the lives and careers of these men, insofar as they impinge upon the career and activities of William Weston Young, and a prime source of information as to the extent of their interaction with him is provided in the handwritten records known as the William Weston Young Diaries (Morton Nance, *The Pottery and Porcelain of Swansea and Nantgarw,* Appendix VIII-B 1942).

Ernest Morton Nance (*The Pottery and Porcelain of Swansea and Nantgarw* 1942, Appendix VIII-B, pages 495 and *ff*.) gives an excellent summary of William Weston Young's Diaries or "*Fact Books*", which number some 26 small paper or leather-bound pocket-sized volumes and two small surveying books, each of the latter containing two months of his Diary (between the dates of March 1st and June 30th, 1821), an *Account of the Salt Works at Droitwich* (January 22nd, 1825–April 17th, 1826) and a housekeeping book for the two periods of October 22nd, 1838–September 9th, 1839 and from January 1st, 1842–February 28th, 1843. Morton Nance is of the opinion that these little books comprise the most important account of the early history of the Nantgarw China Works in existence, as Billingsley and Walker never wrote anything down about their dealings with Young or others with whom they came into contact. Hence, modern historians and researchers do not have the benefit of reference to a Nantgarw pattern book or work books on which some deductions about the contemporary chronology can be made with regard to service commissions, correspondence with clients, comparison with other factories and events.

A complete list of the books which comprise the corpus of the William Weston Young Diaries is provided in Table 3.1, which also illustrates the now missing items and also notes several that seemed to have been in existence (but which are now no longer extant) for William Turner's survey accomplished for his own book on the Swansea and Nantgarw China Works (*The Ceramics of Swansea and Nantgarw* 1897) published some 45 years before that of Morton Nance in 1897. A wealth of information is certainly contained within these Diaries, which is often rather difficult to apprise easily because of the often hastily written and terse comments made in the text, which are also completely devoid of punctuation; nevertheless, some highly important points of information can be assimilated from a study of the data contained within from the point of view especially of establishing a chronology or timeline involving the interactions of the major personalities involved in the Nantgarw China Works scenario.

Although there are some 30 separate books surviving in William Weston Young's diary archive, the first date record begins with October 28th, 1801, and the last with February 28th, 1843, with some absences as noted in Table 3.1. These records therefore encompass the life of William Weston Young between the ages of 25 and 67, when he was at his most productive—and, of course, they cover his major interests in the subject matter of the title of this book: porcelain manufacture and silica brick making. Many of the entries, as indicated by Morton Nance's brief precis, which covers 43 closely spaced pages of text in an Appendix, can be seen as rather mundane (such as details of a supper of bread and cheese taken by Young on an overnight stay in a hostelry in Merthyr) even though these are a reflection of the travels and business dealings of William Weston Young. A digested summary of diary extracts relevant to the topic of our thesis here is as follows.

An entry corresponding to January 30th, 1814, shows that Young paid an account of Walker and Bealey of £30-0s-0d; this was followed by a February 24th, 1814, entry which stated that Young went to Nantgarw and paid Walker and Bealey, £3-3s-0d. This latter entry (Fig. 3.1) is significant in that it records the very first mention of a visit to Nantgarw by Young—although, of course, he could have well already done so to meet with Walker and Billingsley on January 30th as he does not mention the place where he actually met with Walker and Billingsley! There then follows a profusion of entries mentioning either Nantgarw or payments to Walker and Bealey and it is clear that William Weston Young was supporting Walker and Billingsley financially and solely at Nantgarw at this time. For example, on March 4th, 1814, Young writes that he arrived at Nantgarw and paid expenses to Walker and Bealey of £1-18s-0d and again on March 5th there is a record of another payment of two months' rent and expenses for them of £30-0s-0d. An interesting record in the Diary also is a series of payments to Thomas Pardoe at Bristol for unspecified services, but probably involving the decoration of pottery and porcelain, such as the entry for £42-0s-0d on March 8th, 1814. No further mention is made of Nantgarw until May 2nd, 1814, when another payment for an unspecified amount is made to Walker and Bealey and also to Thomas Pardoe at Bristol on the same date for £30-0s-0d. On May

Table 3.1 William Weston Young's Diaries

Book No.	Dates of entries
1	October 28, 1801–February 11, 1802
	February 12, 1802–July 24, 1806
2	July 25, 1806–April 25, 1807
3	April 17, 1807–July 17, 1807
	July 18, 1807–December 31, 1807
4	January 1, 1808–February 29, 1808
5	March 1, 1808–February 2, 1809
6	February 3, 1809–May 29, 1809
7	June 1, 1809–January 18, 1810
	January 19, 1810–August 28, 1810
8	August 29, 1810–February 9, 1811
9	February 10, 1811–April 20, 1811
10	April 21, 1811–May 25, 1811
	May 26, 1811–June 7, 1811
11	June 8, 1811–April 19, 1812,
12	April 20, 1812–October 3, 1812
13	October 5, 1812–April 13, 1813
14	May 1, 1813–August 1, 1813
15	August 23, 1813–March 8, 1814
16	March 10, 1814–December 5, 1814
	December 6, 1814–August 6, 1815
17	August 7, 1815–May 6, 1817
18	May 7, 1817–October 15, 1817
19	October 16, 1817–February 25, 1818
20	February 26, 1818–August 31, 1818
21	September 1, 1818–February 28, 1819
22	March 1, 1819–December 1, 1819
23	December 2, 1819–December 31, 1820
	January 1, 1821–February 28, 1821
24	March 1, 1821–April 30, 1821
25	May 1, 1821–June 30, 1821
	July 21, 1821–January 21, 1825
26	January 22, 1825–April 9, 1825
	April 10, 1825–November 8, 1825
27	November 9, 1825–April 17, 1826
	April 18, 1826–December 1, 1826

(continued)

Table 3.1 (continued)

Book No.	Dates of entries
28	January 1, 1827–July 24, 1827
	July 25, 1827–October 8, 1831
29	October 10, 1831–December 31, 1832
	January 1, 1833–December 31, 1842
30	January 1, 1843–February 28, 1843

Footnote Entries in *italic script* indicate missing volumes which conform to these dates. No diaries exist prior to October 28, 1801 or after February 28, 1843

13th, Young records that he reached Nantgarw and slept there—a longer than usual meeting perhaps that necessitated a stay over? A possible clue for the reason for this extension is given in an entry for May 20th, only a week later, where Young records that he had agreed with George Haynes of the Swansea Pottery, that he should go to Nantgarw and assess the situation there: around this time, Young accepts the offer of a partnership in the Nantgarw China Works with Billingsley and Walker and it seems that he sought the opinion of George Haynes about this proposal and development—Morton Nance believed that Haynes was the only person in South Wales with the relevant experience in china manufacturing who was able to advise Young of the practicalities of china production at Nantgarw. It seems that Haynes, even so, did not personally go to Nantgarw to meet with Billingsley and Walker as expenses of £3-0s-0d were paid by Young to Haynes' representative on June 28th, 1814, a William Baker, who actually visited Nantgarw on his behalf. Obviously, Young's opinion about the setup at Nantgarw was agreeably reinforced by the advice given by Haynes and his representative as he was involved just a little later in a formal partnership with Billingsley and Walker at the Nantgarw China Works—and they became co-signatories of the *Memorial* (reproduced in Chap. 2) officially requesting British Government support for the Nantgarw China Works which was submitted on September 5th, 1814. The manner of address of Billingsley and Walker in Young's Diary is interesting—it is always Walker and Beeley/Bealey/Beely, and never Billingsley and Walker: the variation in the diminutive for Billingsley is appreciated, this even persisting to the official Memorial signature on September 5th, which again has the signatories in the order of Samuel Walker, William Beeley and William Weston Young. William Billingsley, it could be argued from this, took second rank to Walker and for some reason continued to use the diminutive appellation (as shown in the handwritten copy of the signatories to the *Memorial* shown in Fig. 2.2). It is also clear that Young had an ongoing business arrangement with Pardoe, probably for the enamelling of porcelain or pottery which would have been decorated and fired in Pardoe's kilns at Bristol.

We are still faced with the question as to how Young first met Walker and Billingsley to set up the operation at Nantgarw China Works, called Phase I: Kyrle Fletcher (cited in Morton Nance, *The Pottery and Porcelain of Swansea and Nantgarw* 1942, p. 241) has apparently found a record of a meeting between Billingsley and Young

Fig. 3.1 Photograph of the entry from William Weston Young's diaries for January 30th, 1817, which records definitively the very first meeting between William Weston Young, Samuel Walker and William Billingsley at the Nantgarw China Works site. This definitively marks the chronological start of phase II of the Nantgarw China Works establishment

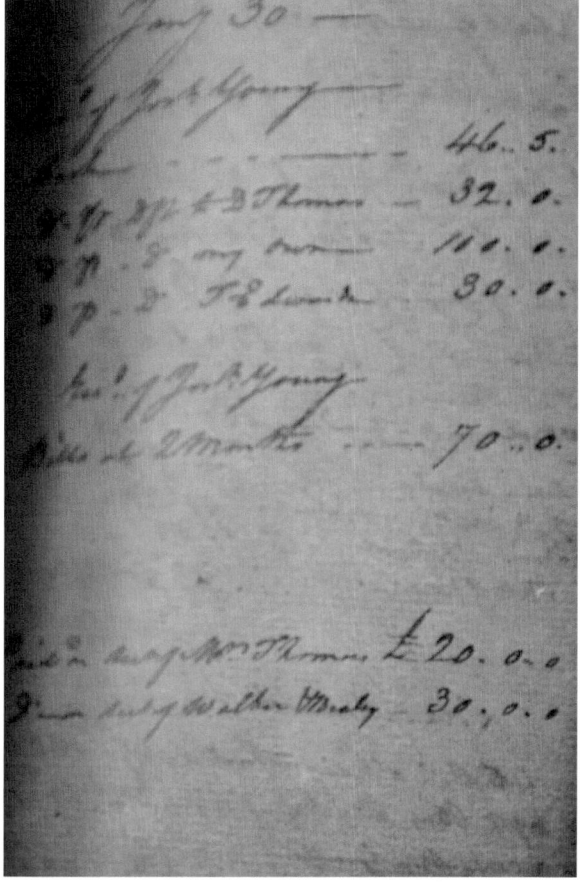

in Margam, which had been arranged by Thomas Pardoe, who, of, course would have worked under Billingsley's tutelage as an apprentice in the Derby China Works and knew Young intimately from their time together at the Swansea Pottery. This is indeed an intriguing prospect, as it would mean that Pardoe had a vital part to play not only in the original setting up of the Nantgarw Phase I operations in 1814 but that he also saw Phase II to completion upon his death there in 1823 as the last incumbent to have tenure at the Nantgarw China Works. Another proposal in the literature is that Billingsley, upon leaving Worcester took the initiative and actually tried to meet with Dillwyn at Swansea, meeting with Young there... but evidence for this is, frankly, rather conjectural as any such meeting was never recorded by Young or by Billingsley.

Much has been made of Young's expenses in financially supporting Billingsley and Walker with numerous entries in the Diary books for payments of small to considerable sums of money between the January and July of 1814—thereafter, presumably heeding Haynes' positive advice, Young became a partner but he still made significant payments of £100 a time to Billingsley and Walker up until the September of 1814, when as a result of their failed application for official Government funding (known as the *Memorial*, see Chap. 2) of the Nantgarw China Works, the operation was doomed to closure because of financial problems. One thing is clear, the £200 awarded to Billingsley and Walker by Martin Barr of Barr, Flight and Barr before their departure from the Royal Worcester China Works was not sufficient for their setting up of Nantgarw Phase I into commercial production—the site was virgin territory for ceramics production and kiln construction alone would have been costly, let alone the purchase of the high quality materials essential for porcelain production and especially the 24 carat gold leaf required for gilding the finished porcelain. On July 2nd, 1814, Young purchased gold from Bristol specifically for decorating Nantgarw china. He spent four days at Nantgarw later in July, between the 25th and 28th June, and it is tempting to argue that this would be when he had formally accepted the invitation of Billingsley and Walker to be partner in the Nantgarw China Works enterprise. It can be estimated that Young gave Billingsley and Walker in excess of £600 personally for the Nantgarw Phase I enterprise, which of course he never recouped and some authors have suggested that the true figure given by Young to Walker and Billingsley is nearer £1000, which would equate to approximately £100,000 today.

On September 17th, 1814, Young records that he met with Billingsley and Pardoe at Margam and Morton Nance has suggested that this presaged a potential move of the Nantgarw operation to Swansea (if Kyrle Fletcher is to be believed then this would have been the second meeting of these three at Margam that year—and neither meeting apparently involved Samuel Walker)—again, the role of Pardoe in these discussions is quite crucial, but where was Walker when these discussions were being undertaken? On September 29th, Lewis Dillwyn is recorded as having met with Young and Walker, the location of this meeting is not stated, but on October 5th, 1814, Dillwyn was at a meeting at Nantgarw with Young (the presence of Billingsley and Walker is not stated) and it is clear that negotiations to secure Billingsley and Walker's relocation to the Swansea Pottery to fulfil Dillwyn's ambition to create porcelain there were now well advanced. Most historical authors have highlighted the role of Sir Joseph Banks, who was the President of the Commission who received the *Memorial* from Walker, Billingsley and Young on September 5th, 1814, whereby arose his informing Lewis Dillwyn of the lack of success of their application and advising him of the opportunity to acquire their special expertise for Dillwyn's porcelain manufacture at Swansea. Dillwyn certainly lost no time in securing his position in this respect and his meeting with Billingsley and Young at Margam just twelve days after their submission of the *Memorial* in London is testament to his keenness to progress his ambition to create porcelain at Swansea. However, we have seen already that George Haynes already knew of the expertise residing at the Nantgarw China Works for porcelain manufacture, as he had been called into advise Young on July 2nd! Haynes was the Manager of the Swansea Pottery at this time, where he had been brought in by the owner Lewis Dillwyn Sr. to run the factory and advise his son, Lewis, who was a complete novice in aspects of the ceramics business.

Apparently, although eager to learn, Lewis Dillwyn was kept very much in the dark by the autocratic Haynes, who ran the business almost independently of Dillwyn Jr. In the absence of Lewis Dillwyn Sr., who was now permanently resident at the family home, Haynes had complete control effectively of the Swansea Pottery. This left Lewis Dillwyn to his own devices and he was wont to wander off and happily pursue his love of the local flora and fauna (as exemplified by his research for his book on the *Confervae* algae, published in 1809) to the detriment of his acquisition of the practical knowledge of running a ceramics factory. Thus, when Sir Joseph Banks alerted Lewis Dillwyn Jr. to the possibilities of taking on Nantgarw personnel the latter seemed to have no prior knowledge of the existence of this operation—but being technically in charge at the Swansea Pottery, Lewis Dillwyn was fully sanctioned to employ Billingsley and Walker when the opportunity eventually arose to achieve his ambition to make porcelain at Swansea.

The employment of Young at Swansea by Dillwyn was not taken up at this stage in 1814, and it transpires that Young was left with the derelict company at Nantgarw, minus Walker and Billingsley, for which he accepted responsibility for its assets and debts, being then declared a bankrupt for the second time (reference to the notice in the *London Gazette*, Saturday, December 17th, 1814); he was required to present himself at the Rummer Tavern in Bristol to hear the bankruptcy proceedings brought against him by his creditors on January 5th, 6th and 28th, 1815. Little more can be gleaned about Young and his Nantgarw role after this entry in his Diaries but an interesting statement reveals that he was now dealing in bricks at Neath Abbey—actually, Young's interest in bricks started at an even earlier stage, as there was a mention of it as early as 1808 in his Diary entries. For example, on February 6th, 1809, Young transported 13,000 bricks to Sir John Nicholl at Merthyr Mawr for £15-0s-0d. Already, it seems that he was selling a type of "firebrick" for furnaces too, whatever that comprised in its raw material composition …!

The transition of Billingsley and Walker to Swansea in October 1814, was itself not as smooth as it appears because Billingsley is recorded in the Diary as returning to Nantgarw "temporarily" on November 5th, 1814, whereas Walker remained at Swansea. There has been some conjecture about this event and there is some credibility to be given to the idea that Dillwyn was trying to emulate the production of a variant of the Nantgarw porcelain body at Swansea to match in with his new kiln construction. It is known that later, in 1815–1817 he did undertake experimental trials with new concepts in porcelain paste (for which we have excellent quantitative data cited in Dillwyn's notebooks (see Appendix B) which are now held in the archive in the Victoria and Albert Museum, London (Dillwyn 1922), assisted by Samuel Walker, whom he had appointed as his head of china manufacture at Swansea. So, we could hypothesise that perhaps Billingsley regarded the earlier attempt in 1814 as a rejection of his ambitions and plans and had decided to remove himself elsewhere temporarily, namely Nantgarw? By this time the Nantgarw kilns had been shut down anyway as then Young was formally responsible for the Nantgarw China Works and there would have been little for Billingsley to do there … unless there were items of already glazed china in the white requiring decoration with the limited supplies of still available pigment materials at his disposal. Whatever the reason for his temporary departure from Swansea, Billingsley soon returned, as vouchsafed by Young later in November: in the meantime, Dillwyn had now received his "threatening" letter from Flight, Barr AND Barr (a copy of the letter sent to Samuel Walker at the Swansea China Works by Flight, Barr & Barr is given in Appendix A) at the Royal Worcester China Works cautioning him about employing Billingsley and Walker at Swansea—but the chronology definitely dictates that this letter could not have activated Billingsley's departure temporarily from Dillwyn's employment and his removal to Nantgarw for a potential set-up there. Perhaps, the use of the Nantgarw moulds initially at Swansea for the immediate production of a new type of porcelain would have been a source of concern for Billingsley, but in these early days the fledgling emergent Swansea China Works would probably have found it difficult to create their own moulds in such a small time scale: Morton Nance (*The Pottery and Porcelain of Swansea and Nantgarw* 1942, p. 266) considers this aspect very carefully and points out that the few surviving examples of Swansea porcelain from this early period of production are very closely similar indeed to their Nantgarw analogues but with a very slight differential measurement overall which can be attributed to shrinkage of the new Swansea paste, ascribed to a greater contraction experienced upon firing. This surely implies that the novel Swansea paste could not have been identical with the Nantgarw version as there was an observed differential coefficient of expansion upon firing. Morton Nance also comments that this early Swansea paste was remarked upon by Dillwyn as being less granular than the Nantgarw version and that the glaze did not penetrate as well. The clear implication is that Billingsley and Walker had not revealed their Nantgarw formulation to Lewis Dillwyn upon joining him at Swansea, yet Dillwyn must have been aware of some of the components of the Nantgarw body at this stage?

After his assumption of the responsibility for the now defunct Nantgarw China Works, the bankrupted William Weston Young turned his attention to other pursuits and interests to earn some money, as evidenced by the multiple entries in his Diary referring to his other activities: he remarks particularly on the success of his wreck salvaging business and it is clear that the two specified wrecks highlighted earlier in Chap. 2, namely the *Anne & Therese* and the *Trelawny,* are but two of those that he successfully raised: in most cases he just refers in his Diary entries to a rather anonymous "wreck", but a noted later exception is the *William Penn,* a schooner lost in the Bristol Channel off Clevedon in Somerset on January 21st, 1820. Young raised her, with much credit accruing and favourable public acknowledgement of his prowess, as this was the first vessel ever to be raised from this part of the seaway, and he was applauded for this feat in *The Cambrian* issue of April 8th, 1820. There is no doubt that Young's wreck salvaging activities were instrumentally significant in enabling him to clear his debts: the *Anne & Therese* cargo of copper ingots mentioned earlier completely cleared the debts of his first bankruptcy and thereby facilitated his ability to finance the early efforts of Billingsley and Walker in establishing Phase I of the Nantgarw China Works to the tune of £600, or perhaps even more. The salvaging of corn from the *Trelawny* assisted him in helping his elder brother Joseph Young pay off significant debts from his Bristol factoring business which had reached trouble: Young raised much of the 3000 bushels of corn cargo on this vessel, which he sold on to his brother for 1s 6d per bushel, which his brother could then sell on at the reduced rate of 8 s per bushel against the going commercial rate of 11s per bushel. In this way, Joseph Young cleared his outstanding debts of £1000, entirely due to his brother's generosity. A rider to the generosity of William Weston Young in helping distressed family is given by Turner (*The Ceramics of Swansea and Nantgarw* 1897, p. 151) who compares him with the "*Admirable Crichton*": this refers to a character invented by J. M. Barrie in a story about the aristocratic family of Lord Loam which became marooned on a desert island with their butler, James Crichton, who immediately adopted the lead in helping the family survive their ordeal and protecting them from hardship, only to revert to his former subservient role upon rescue without any expectation of gratitude.

Between January, 1815, and January, 1817, after the collapse of the Nantgarw China Works enterprise following the unsuccessful bid for financial support from the government, William Weston Young turned to other activities and Nantgarw is not mentioned at all in his Diaries. His surveying business obviously took him through the county of Glamorgan, where he was engaged in drawing up plans and surveying land for some wealthy patrons, using his artistic drawing skills also to good effect in an appointment at the Cowbridge Free School, where he taught up to four pupils weekly for several years. He was obviously very drawn to this aspect and he describes his meetings at Cowbridge, which often occurred on the way to and from other meetings in the County. The first record of this appears on August 13th, 1817, and he regularly notes this activity, frequently occurring on a weekly basis until 1821, even naming his drawing pupils accordingly.

A very important record appears for January 12th, 1817, where Young first meets with William Billingsley since the folding of Phase I of the Nantgarw China Works at his home in Newton Nottage. This is significant as Samuel Walker is not involved in this meeting: Billingsley had by then already parted company for the second time with Lewis Dillwyn, who had still retained the services of Samuel Walker for his experiments in trying to create a more robust version of his Swansea duck-egg porcelain which was held in high esteem by a discerning clientele, but which, like its Nantgarw analogue, suffered high kiln losses in firing. There is much to support the idea that the generation of the new Swansea *trident* ware (Edwards, *Swansea and Nantgarw Porcelain: A Scientific Reappraisal* 2017), containing a significantly higher proportion of soaprock (steatite, magnesium silicate), precipitated this final departure of William Billingsley from Swansea as he must have viewed this production as a seriously retrograde step in his own ambition to create the world's most translucent porcelain—which, unfortunately for the eventual survival of the Swansea China Works, was an assessment echoed by the potential clientele. We can theorise about the purpose of this meeting between Young and Billingsley at Young's home in Newton Nottage but it would be in accord with later events that the re-opening of the Nantgarw China Works was high on the agenda. It is significant that Walker was not present at this meeting and we can hypothesise that perhaps Billingsley did not wish to alert Dillwyn to his future plans, which necessarily involved input from Walker as the expert on kiln construction and technology. Billingsley may even have been rather aggrieved that Walker had elected to remain at Swansea to assist in Dillwyn's experiments and had not left with him in December 1816. Young records his visits to the Nantgarw site subsequently on April 17th and May 1st, 1817, and it is tempting to propose that he was then seriously assessing the proposition for its reopening, as would have perhaps been suggested by Billingsley at their meeting earlier in the year. In June of 1817, several entries in his Diary book suggest that Young was now actively seeking financial support from local landowners and wealthy patrons for the re-creation of Nantgarw porcelain—but his other interests still pervaded, for example, the sale of bricks occupied him constantly and completely between July 22nd and 28th, 1817! On August 13th, he teaches three pupils drawing skills at the Cowbridge Free School on the way to paying Walker and Billingsley some expenses at Nantgarw—so their differences now reconciled, the kiln manager and master porcelain producer and decorator were now clearly reunited again at Nantgarw for potentially the opening of Phase II of the Nantgarw China Works. This chronology is itself interesting as it confirms that Walker was still in touch with Billingsley at Nantgarw before he actually left Swansea in September 1817. The visits of Young to Nantgarw then continue steadily until November 1817, when Walker is now ensconced there, various payments being recorded being made from Young to Walker and Billingsley—who is still referred to as Bealey in the Diary records. On January 6th, 1818, the final payment was recorded from Young to Walker and Billingsley of £20-0s-0d was made at Nantgarw after which the most welcome support from "*the ten gentlemen of Glamorgan*" was received (Edwards, *Swansea and Nantgarw Porcelain: A Scientific Reappraisal* 2017) and payments then were made to Billingsley and Walker from this new source. We should not forget that up until this time William Weston Young was still an undischarged bankrupt from taking on the debts of the Nantgarw China Works Phase I operation—yet he was able to support Walker and Billingsley financially and solely from his own funds during the start-up of Phase II at Nantgarw until the £1000 awarded by the ten sponsors, who each contributed £100, was actioned and running. The bankruptcy issue possibly explains why the actual collection of the financial support for this phase of the Nantgarw China Works was not overseen by William Weston Young personally—as an undischarged bankrupt he would probably have been legally liable to handing over any monies he collected from this activity to his creditors—hence, The Hon. William Booth Grey of Duffryn St Nicholas was nominated to collect the money. It is also clear that William Weston Young's personal activities as a land surveyor over the preceding two years or so enabled his acquisition of a wealthy clientele who could assist him materially in this endeavour. Morton Nance (Morton Nance, *The Pottery and Porcelain of Swansea and Nantgarw* 1942, p. 363) has researched much relevant extant documentation regarding the patronage of the Nantgarw China Works in its start-up in 1817 and has discovered a list of patrons (which he has termed *subscribers*, after William Weston Young's own terminology), whom he has listed as follows: William Weston Young of Nantgarw; The Hon. William Booth Grey of Duffryn St. Nicholas, High Sheriff of Glamorgan; John Crichton Stuart, The Marquess of Bute, Cardiff Castle; Sir John Nicholl, Merthyr Mawr; Robert Jones, Fonmon Castle; Wyndham Lewis, MP, Green Meadow, Llanishen; Dr Whitlock Nicholl, Cowbridge; The Mansell Talbots of Margam Abbey; Griffith Llewellyn, Baglan Hall; The Rev. William Perkin M. Lisle, LL.D, Prebendary of Llandaff and

Rector of St Fagan's; Lord Windsor, St Fagan's; R. Franklin Esq.; Thomas Wyndham, Dunraven Castle, was initially approached and he reacted favourably to the invitation with an offer of support but he died before his support could be realised. A simple calculation reveals that there are twelve names associated with this list—but if we exclude William Weston Young on account of his bankrupt status and William Booth Grey as the coordinator and collector, then the others probably represent the ten sponsors discovered by Morton Nance and referred to in the earlier documentation. In addition, as was indicated above, Young had several influential business friends and contacts through his surveying profession who probably also were persuaded to support the financial re-start of the Nantgarw factory, including: Walter Coffin, Llandaff Court; John Bruce, Duffryn Aberdare; William Crawshay, Cyfarthfa Castle, Merthyr; Crawshay Bailey, Merthyr; Richard Hill, Plymouth Lodge, Merthyr; John Edwards, Rheola House, Vale of Neath; William Forman, Penydarren, Merthyr Tydfil; Richard Blackmore, Melyn Griffith; J.J. Guest, Dowlais Ironworks; T.B. Rous, Courtyrala; Thomas Edmondes, Cowbridge; R.H. Jenkins, Llanharan House, Llantwit Fardre; William Williams, Aberpergwm House, Vale of Neath; The Rev. John Traherne, St Hilary, Cowbridge; William Vaughan, Llantrisant and Edward Edmunds, Penyrhos, Nantgarw (landlord of the Nantgarw factory site). It is known that several of these purchased large services later from the Nantgarw factory which were painted locally, such as Sir John Nicholl of Merthyr Mawr, William Williams of Aberpergwm House and Wyndham Lewis of Green Meadow, Llanishen. Two dessert plates from the superbly decorated service made for Wyndham Lewis are shown in Fig. 3.2, decorated locally by Thomas Pardoe (John et al. *Nantgarw Porcelain Album* 1975). In total, the relatively handsome sum of £2100 was raised, including William Weston Young's personal contribution of about £1100 to start up porcelain production at Nantgarw in the second phase in 1817 which can be compared with the modest sum of £200, with a further contribution from Young, was invested by Billingsley and Walker in their first phase Nantgarw operation in 1814, this being their severance payment received from Barr, Flight & Barr, of the Royal China Works at Worcester.

On March 18th, 1818, an important record in the Diaries reveals that his creditors met with Young in Bristol to discharge his debts and absolve his bankruptcy: the *Glocester Journal* issue of March 7th, 1818, notified creditors of this meeting, which must have been a huge relief for William Weston Young. By this stage, however, despite its successful launch and an insatiable demand for its china which must have exceeded expectations, the Nantgarw factory was already starting to feel the consequences of its appallingly high kiln wastage and an inability to meet these demands in production was setting in.

Young appears not to have taken much interest in Nantgarw during the year 1818, leaving the running of the factory to Billingsley and Walker—instead he delighted in recording his weekly meetings with his drawing pupils in Cowbridge throughout that year and into 1819, with some unspecified wreck-raising activity being interspersed with his teaching. Having gone out on a limb with his creditors and sponsoring Billingsley and Walker earlier in the year to enable them to set up Phase II of the Nantgarw China Works, which was an immediate success, Young seemed content to adopt what was a very much a backstage role for the ensuing months.

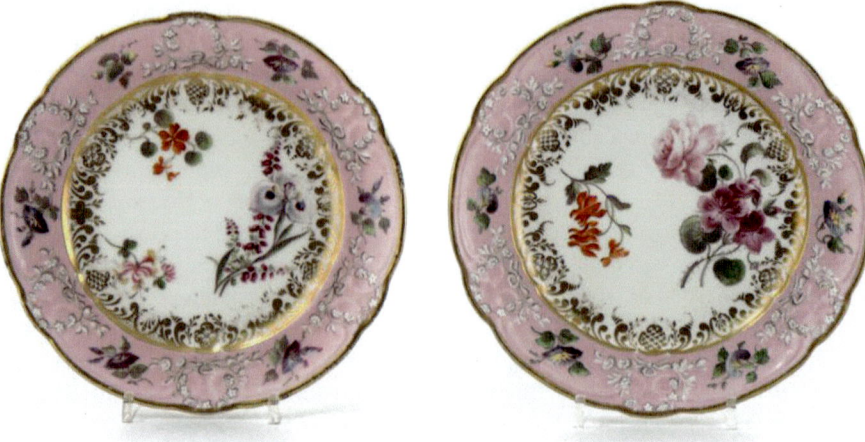

Fig. 3.2 Pair of plates from the esteemed Nantgarw dessert service prepared for Wyndham Lewis MP, of Green Meadow, Llanishen, ca. 1821–1822 by Young and Pardoe. This service was part of the remaining stock left at Nantgarw after the departure of Walker and Billingsley for Coalport in 1820: it was decorated by Thomas Pardoe with sprays of garden flowers, with profuse gilding and a very attractive pink rose pompadour ground colour at the verge. It will have applied the Nantgarw no. 2 final glaze perfected by Young and Pardoe. Private collection

The success of the new Nantgarw operation was phenomenal and their sole agent in London, John Mortlock of Oxford Street, informed Billingsley that he could take all the porcelain they could produce in the white to decorate in their own atelier and sell to a demanding clientele who were prepared to purchase the porcelain at significantly increased premiums, rivalling the much-vaunted and desirable Sevres porcelain (Edwards, *Swansea and Nantgarw Porcelain: A Scientific Reappraisal* 2017). It was even suggested by perhaps envious competitors that Mortlocks passed off Nantgarw porcelain, with an appropriately enhanced premium, as the very desirable Sevres, which perhaps was made possible by the Nantgarw practice of only incorporating their impressed mark on flatware such as plates—other items such as tea and coffee services, spill vases and speciality pieces such as cabaret sets, toilet sets and cabinet cups were not so marked, and dishes were only rarely so.

Several interesting historical points outside of the porcelain field and worthy of note emerge from a study of William Weston Young's diary book entries: firstly, on October 2nd, 1819, Young surveyed land for a proposed *"Rail Way"* from Hirwaun to Pont Walby—this was chronologically a significant time before the commencement of the railway era in Great Britain, which began with the Stockton to Darlington run and the steam locomotives, *Puffing Billy* and *Locomotion* No. 1, followed closely by the Rainhill Trials in 1829, a speed and endurance test for steam locomotives which was won by George and Robert Stephenson's *Rocket* sustaining an average speed of 30 mph over five runs. Of course, these earlier "rail ways" mentioned by Young probably referred to horse-drawn vehicles and trams on tracks such as that built locally at Penydarren near Hirwaun for iron and coal haulage by the Cornishman, Richard Trevithick in 1803. Young published a map of his rail way survey, which he referred to in entries on November 11th and 12th. Secondly, Young's interest in bricks was fostered throughout his diary, and many entries refer to the carriage and sale of this commodity—a particularly relevant entry relates to August 7th, 1820, when he comments on his engagement with William Bedford, who had constructed and operated the muffle furnace in Young's home at Newton Nottage, to make bricks: so, from selling and carting bricks in the earlier days, Young had now moved on to manufacturing them as well. This will be especially relevant to our considerations of his move into his manufacture of the specialised furnace silica firebrick just some year or two later as Young was obviously experienced and acquainted in the manufacture of normal bricks at the outset.

Billingsley left Nantgarw with Walker in April 1820, and on September 17th, 1820, Young opened negotiations with Edward Edmunds, the landlord of the Nantgarw China Works site, to lease the property in his own name. On October 24th, 1820, Young employed William Bedford (his associate in the brick-making enterprise, as cited above) to construct a muffle furnace at Nantgarw for "*decorating china*": clearly, this was Young's first move in his plan to re-open the Nantgarw China Works for business, but he was still selling bricks some two days later, in a Diary entry on October 26th!

William Weston Young was now dividing his time between the emerging china business in Nantgarw and his brick-making business, for between December 15th and 19th, 1820; Young was at Dinas in the Vale of Neath, where he was proposing to establish his new venture in silica brick manufacture. He also still required a decorator for his china stock at Nantgarw and we now see Thomas Pardoe being targeted for this purpose: on February 12th, 1821, Pardoe was engaged by Young to paint at Nantgarw and he arrived there in residence on February 20th—an entry in the Diary on February 21st, 1821, indicates that the glazing kiln, or glost kiln, was fired up at Nantgarw for the first time. On the 2nd March 1821, Young refers to Thomas and William Henry Pardoe being at Nantgarw firing up the muffle furnace with his new glaze frit, the later so-called Nantgarw No. 2 glaze. Young's hiring of several named china manufacturing people from the old Phase II production at Nantgarw, such as Thomas Pole and David Jones, in April and May of the same year, can only mean that this is when Young had decided that he was going to make porcelain at Nantgarw again. The Diary is now full of entries noting Young's meetings with the workforce at Nantgarw and it is likely that he was now experimenting in the manufacture of a trial porcelain formulation there at this time. Unfortunately, several volumes of his Diary between late 1821 and 1825 are now missing so it is unclear when the experiments were commenced and if any note was ever made about their successful conclusion or otherwise. Young's notes from October 19th, 1822, referred to his spending some time at Dinas and his leasing the site from the Marquess of Bute, and the manufacture of silica refractory bricks there was now clearly a realistic proposition.

From the winter of 1829–1830, William Weston Young was living in "Fairyland", Neath, whence he moved again on April 19th, 1832, to a house in Neath incorporating a shop, store and a large garden with outbuildings. Morton Nance has collated much information to show that Young now spent most of his time at the Dinas Firebrick Works and also in a local colliery in the Vale of Neath, along with land surveying and painting pottery and porcelain. This colliery was probably the Gnoll Colliery, which has been mentioned earlier in Chap. 2 as a "failed enterprise". On April 21st, 1832, he is recorded as having taken four trial china clay specimens to the Glamorgan Pottery (*sic.*, here mis-named the Swansea Pottery, which had actually folded after Dillwyn sold out to the Bevingtons in 1820) for firing, and he went again on May 2nd to retrieve the fired samples from Martin Bevan (the proprietors of the Glamorgan Pottery were Baker, Bevan and Irwin). Young commented that on May 21st he called again to collect some trial pieces which "*had not fir'd up the Heat of their Biscuit Kiln not sufficient for this body...*" Throughout May and June and through July, he persisted with these trials and took specimens to the Glamorgan Pottery as well as using his own muffle furnace at Newton Nottage to decorate ceramics, recording the fact that he had also acquired a considerable quantity of earthenware from the Glamorgan Pottery for decoration and glazing—for example, a September 3rd Diary entry states that " *Rec'd ware from Swansea—22 Flats. 5 Large. 5 Middle. 12 Small 18 Match Potts—2 Setts of Tea Ware*".

It seems from the Diary entries that William Weston Young wrote to Apsley Pellatt of the London Porcelain Co. that his experimental products fired in the Staffordshire Potteries at Machin and at Minton had resulted in "*a very superior fine china*". Even as late as October 19th, 1838, Young was still experimenting with trials in china manufacture. As will be seen, Young and John Wright of Bristol were busy creating a market around this time for the sale of a ready-prepared *Nantgarw dry mix* porcelain substitute, which was probably based on Young's hard paste formulation recipe: this never really succeeded with potential purchasers in the Staffordshire Potteries, including Machin and Minton, who must have expected a better Nantgarw-like body than was actually achieved by the addition of water to the *dry mix* paste preparation. What must at first have appeared to have been an attractive proposition to the potential china manufacturer purchasers was that Young and Wright were offering a ready-mix dry porcelain powder for £15 per ton. Calculations reveal that the proportionate base costs for a typical dinner service and a tea set would be in the region of about £1 10s and 10s, respectively, which would indeed be a worthwhile investment if the resulting china had anything like the translucency of the genuine Nantgarw porcelain. It is recorded that several small deliveries of *dry mix* were negotiated but the purchasers were clearly disappointed with the result.

The next chapters will provide some details of the personalities who interacted with William Weston Young and whose influence had bearing upon his activities and decisions: Thomas Pardoe, Samuel Walker, Lewis Dillwyn and William Billingsley.

References

L.W. Dillwyn, *British Conferva or Coloured Figures and Description of British Plants Referred by Botanists to the Genus Conferva* (W. Phillips. George Yard, Lombard St., London, 1809)

L.W. Dillwyn, *Notes on the Experimental Production of Swansea Porcelain Bodies and Glazes. Made by Lewis Weston Dillwyn with Samuel Walker at the Swansea China Works Between 1815 and 1817. Presented to the Library of the Victoria &Albert Museum, South Kensington, London by John Campbell in 1920. Reproduced in Eccles & Rackham, *Analysed Specimens of English Porcelain*, 1922, see reference below

H.G.M. Edwards, *Swansea and Nantgarw Porcelain: A Scientific Reappraisal* (Springer, Dordrecht, The Netherlands, 2017)

W.D. John, G.J. Coombes, K. Coombes, *The Nantgarw Porcelain Album* (Ceramic Book Co., Newport, 1975)

E. Morton Nance, *The Pottery and Porcelain of Swansea and Nantgarw* (Batsford, London, 1942)

W. Turner, *The Ceramics of Swansea and Nantgarw* (Bemrose & Sons, Old Bailey, London, 1897)

W.W. Young, *Original Manuscript Fact Books* (Glamorgan Records Office, Cardiff, 1801–1843)

Chapter 4
Thomas Pardoe

Abstract An account of William Weston Young's interaction with Pardoe from their early days at the Swansea Pottery to their final partnership at the Nantgarw China Works.

Keywords Pardoe · Young · Ceramic decoration · Nantgarw China Works · Local decoration

Born 3rd July, 1770, in Derby, Thomas Pardoe was apprenticed in the 1780s at Derby China Works, where presumably he would have first encountered William Billingsley, then a rising star in the enamelling workshop there under the tutelage of Edward Withers, where Billingsley performed his first major artwork on the Prince of Wales service, pattern No. 65, commissioned by Prince George, The Prince of Wales, in 1786 (Edwards, *Swansea and Nantgarw Porcelain: A Scientific Reappraisal* 2017). William Billingsley succeeded Edward Withers as chief decorator in the Derby enamelling workshop in 1790, where he remained until 1795. Although it seems incomprehensible that William Billingsley as chief decorator and in charge of the enamelling workshop would not have been aware of Thomas Pardoe who worked in the same unit, there is no historical evidence that they actually did meet. Before Billingsley's departure from Derby in 1795, Pardoe moved to Worcester and then on to the Swansea Pottery where he decorated creamware between 1795 and 1809, meeting up there with William Weston Young and Lewis Dillwyn. Several important items of earthenware decorated by Pardoe have surfaced from his time at the Swansea Pottery including an important supper set decorated with animals in vignettes on a pale blue enamelled background by both Pardoe and Young and ascribed with a date of 1806.

Williams (*Inaugural Lecture to the Nantgarw China Works Trust* 1993) has commented that Pardoe's painting at the Swansea Pottery was strictly limited to botanical subjects with little opportunity to broaden out into flower painting and that this eventually might have prompted his move to Bristol in 1809, where he could explore and develop his wider talents in this direction. It could also be an explanation for his move finally from Bristol to the Nantgarw China Works in 1821, where he would presumably have had a free rein from William Weston Young in amplifying and demonstrating his decorating skills on the highest quality porcelains.

© Springer Nature Switzerland AG 2019

H. G. M. Edwards, *Porcelain to Silica Bricks*,

https://doi.org/10.1007/978-3-030-10573-0_4

Fig. 4.1 A Nantgarw
porcelain coffee cup with
typical heart-shaped handle,
from the *Spence-Thomas*
breakfast service, decorated
locally by Thomas Pardoe,
ca. 1821–1823. Private
collection

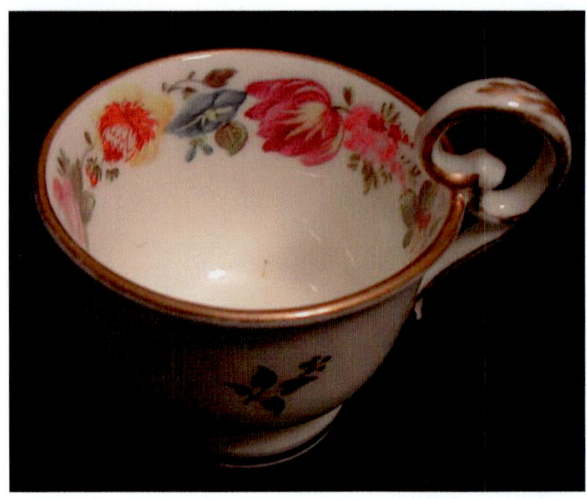

In 1809, Thomas Pardoe moved from Swansea to Bristol where he established
a business and a reputation for the quality decoration of porcelain and ceramics
bought in from other factories, such as Coalport. Much of his work there centred on
floral decoration. He remained there until 1821 when he was approached by William
Weston Young to decorate the remaining stock of porcelain at Nantgarw: for the first
few months it appears that he oscillated between Nantgarw and his business at Bristol
but then he set up permanently at Nantgarw to concentrate upon his enamelling work
there. He died in July 1823, whilst still decorating this remaining stock of Nantgarw
porcelain for "local consumption" such as the *Spence-Thomas* breakfast service,
having been engaged by Young to do this—using the special new glaze that Young
had perfected, the so-called Nantgarw No. 2 glaze. Pardoe's work on Nantgarw
porcelain is highly prized by collectors today: an example is the coffee can from the
Spence-Thomas service shown in Fig. 4.1, with its superb garland of garden flowers
at the rim (Edwards, *Swansea and Nantgarw Porcelain: A Scientific Reappraisal*
2017).

It is still a matter of conjecture, according to some historians, as to whether or not
Young made porcelain at the Nantgarw site from 1820, which Pardoe then decorated:
several authors have dismissed this idea whereas others think it possible on account
of the empirical experiments that were carried out and documented by Young at
Nantgarw between 1821 and 1822 in his Diaries to create a new porcelain in an
attempt to generate funds to ease his precarious financial predicament at the Nantgarw
China Works. William Weston Young does record in his diary entry for October 19th,
1819, that he commenced to undertake trial experiments in porcelain manufacture: the
details of his composition and formulation are given in Table 4.1 and will be discussed
later. The formulations in the recipes are given in a code and it is not a simple matter
to transpose these into the raw materials for comparison with other formulations,
but a realistic key to this code is provided in Table 4.1. As will be seen later, from

Table 4.1 Experimental porcelain paste trials of William Weston Young, October 19th, 1819, with identification of his coded raw materials: FO, V, NC, B, HC, K and L

Number	FO	V	NC	B	HC	K	L
No. 1		120	30	8	3	1	
No. 2		120	30	8	2	2	
No. 3		120	30	8	1	3	
No. 4		120	30	8		4	
To 100 parts of each of the above when fritted and ground add 50 parts B							
No. 5		120	15	4			8
To 100 parts of the above when fritted and ground add 54 parts B, 30 parts N and 2 parts K							
No. 6		120	15	4			8
To 100 parts of the above when fritted and ground add 54 parts B and 30 parts N							
No. 7		60	120	60			
No. 8	120		240	120			
No. 9	70		120	70			
No. 10	70		120	70		2	

Code Deduction
V Sand; *NC* Norden clay; *B* St. Stephen's clay; *HC* Borax; *FO* China stone/Soaprock; *K* Potash/Pearl ash; *N* Nitre; *L* Lead glass/Lime

the recipe compositions of his experiments Young clearly did not have access to the Nantgarw porcelain formulation used by Billingsley and Walker and what he was attempting to make really constituted a hard paste porcelain, incorporating soaprock but not bone ash. Hence, we can conclude, as have others, that William Weston Young actually commissioned Thomas Pardoe to decorate what was effectively the Billingsley/Walker stock of porcelain in the white remaining at Nantgarw, some of which would have been left unglazed in biscuit porcelain. The latter would have then had applied the new Young Nantgarw No. 2 glaze discussed above. Some accounts have queried the amount of porcelain remaining in stock at Nantgarw, especially since John Mortlock, the London agent, is on record as demanding all that Billingsley could produce from his kilns and have suggested that it would not have been sufficient to engage the attention of Pardoe in a full-time decoration schedule for two years from 1821 until 1823. However, it must be remembered that for the Billingsley production period of the preceding two years from 1817 until 1819/20, only the best and perfect specimens of Nantgarw porcelain would have been sent to Mortlock's in London—representing at most only about 10% of the total output from the kilns—and much of the remainder would have been deemed unsatisfactory for that purpose, and perhaps a large amount of this being unusable at all because of warping and sagging defects. Of this, it is certain that some items were decorated locally by Billingsley and others at the factory, and examples of these would have included the *Homfray* service, whilst other wares would have been judged to have been slightly defective or too

(a)   **(b)**

(c)

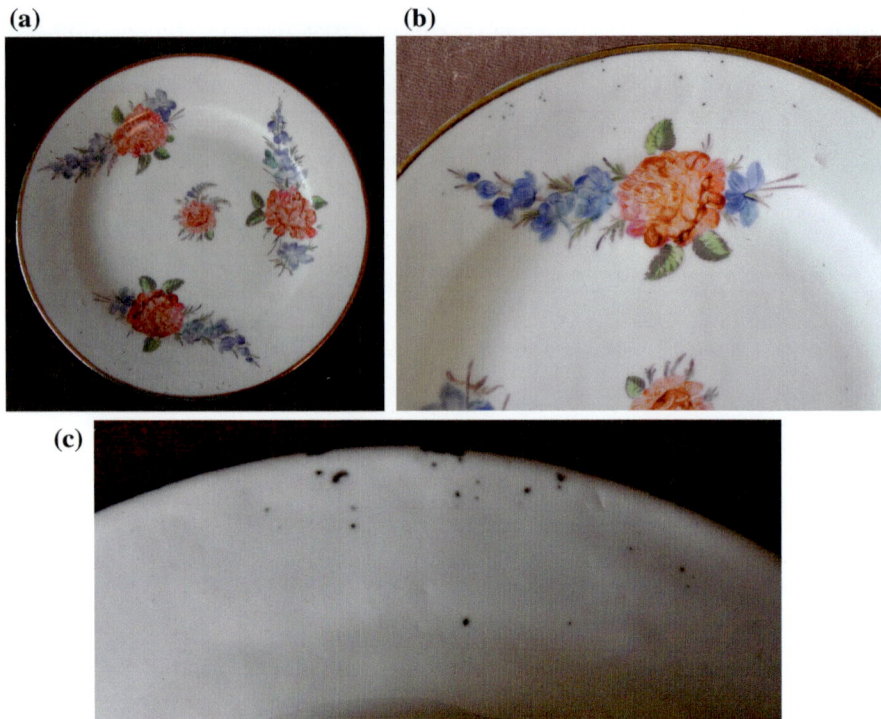

Fig. 4.2 a–c Nantgarw porcelain dessert plate, locally decorated with roses and delphiniums and with a single band of gilding at the edge, Billingsley-Walker period glaze, ca. 1817–1819. Several blemishes mark this plate as being not suitable primarily for the London market, including several black spots arising from carbon particles blown onto the glaze in the glost kiln, which can be seen in the close-up views of the reverse of the plate in (**c**). It can be concluded that this plate was sold locally and decorated by possibly May Hewitt, Lavinia Billingsley before she died in September 1817 or perhaps some other unnamed junior hand. Private collection

warped for anything but a simpler decoration to have been applied but still saleable; an example of this would be the dessert plate shown in Figs. 4.2a–c, which has several dark, localised blemishes caused by carbon particles adhering to the paste during firing and a slight warping of the circular edge—these would have probably have negated its passing through to John Mortlock for decoration and subsequent sale in London but nevertheless it is still an excellent example of locally decorated Nantgarw porcelain and the translucency is still superb. This plate has been decorated simply with sprays of roses and delphiniums, probably by someone such as May Hewitt, or even perhaps Lavinia Billingsley, and sold on for local consumption: examination of this Nantgarw plate reveals the impressed NANTGARW C.W. mark, the Billingsley porcelain paste and the white Billingsley glaze, along with a superb translucency. It is a typical Nantgarw locally-decorated product from the Billingsley/Walker era and was considered suitable for the application of expensive edge gilding in gold leaf, which was not common during the Young and Pardoe period, when the expense of gilding negated its specific usage—hence, edge enamelling in blue, green and chocolate coloured enamels was preferred.

This author believes that the remaining stock at Nantgarw after the departure of Billingsley and Walker was quite considerable and that many items that had previously been condemned as being unsuitable for the London saleroom were decorated and offered for sale and, of these, most would have been salvageable in this way by Young and Pardoe. Of course, it is to be expected that many items from the kiln firing processes would also have been destroyed by Billingsley and these would constitute the major proportion of the waste dump at Nantgarw including particularly the heavily "sagged" porcelain, or overfired items, and those which had suffered irredeemable cracks and fissures during the firing. Some examples of broken shards recovered from the Nantgarw waste dump are shown in Figs. 4.3a–d which exhibit these characteristics. Today in the auction rooms, Nantgarw porcelain items are being offered for sale with firing cracks and blemishes, sometimes filled in with glaze, which indicates that ware which in other factories might have constituted slight "seconds" were actually decorated and sold locally to partially recoup the losses in firing.

Thomas Pardoe's son, William Henry Pardoe, who was born in Swansea in 1803, took over the derelict Nantgarw China Works in 1833, after ten years of closure, renaming it the Old Nantgarw China Works, where he made stoneware bottles and a brown glazed earthenware of the Brameld-type called "*Rockingham*", then moving into clay pipe production of 10,000 pipes per week, until the invention of ready-rolled cigarettes by the US Tobacco company sounded the eventual death-knell of clay pipes world-wide in the late 1880s.

Recent documentary evidence attests that William Henry Pardoe also attempted to make porcelain at the Nantgarw works site: this came about through the offer of support for re-starting porcelain production at Nantgarw in about 1834 from the Marquess of Bute in the form of a £1000 investment, which according to Pardoe's wife, Mary, was rejected outright. However, Pardoe's earthenware business was thriving in Nantgarw by the 1840s and, like his father and afterwards William Weston Young, he then turned to Bristol, where he established a new pottery for clay pipe production in Avon St, St. Philip's, Bristol. Actually, Pardoe took over the old Bristol Pottery which had been run by Edward Yabbicom and Henry Yabbicom until it closed in 1844. Then, William Henry Pardoe announced that he would soon commence the production of porcelain again at Nantgarw which would be financed by the auction of horse-drawn carriages at Nantgarw on Monday and Tuesday, the 15th and 16th August 1843. This idea did not reach fruition at that time and was further delayed as William Henry's eldest son, William, died at the age of thirteen of an unspecified illness. Meanwhile, his clay pipe manufacturing business was thriving and the 1851 census records show that 18 pipe makers were employed in his factory at Nantgarw and 25 at Bristol. A survey of clay pipemakers in Wales from the Census returns provides an intriguing list of names of the workforce (Evans and Jones 1981) which includes Timothy Bevington, late of the Swansea China Works, operating out of the Hafod Pottery in Swansea in 1830.

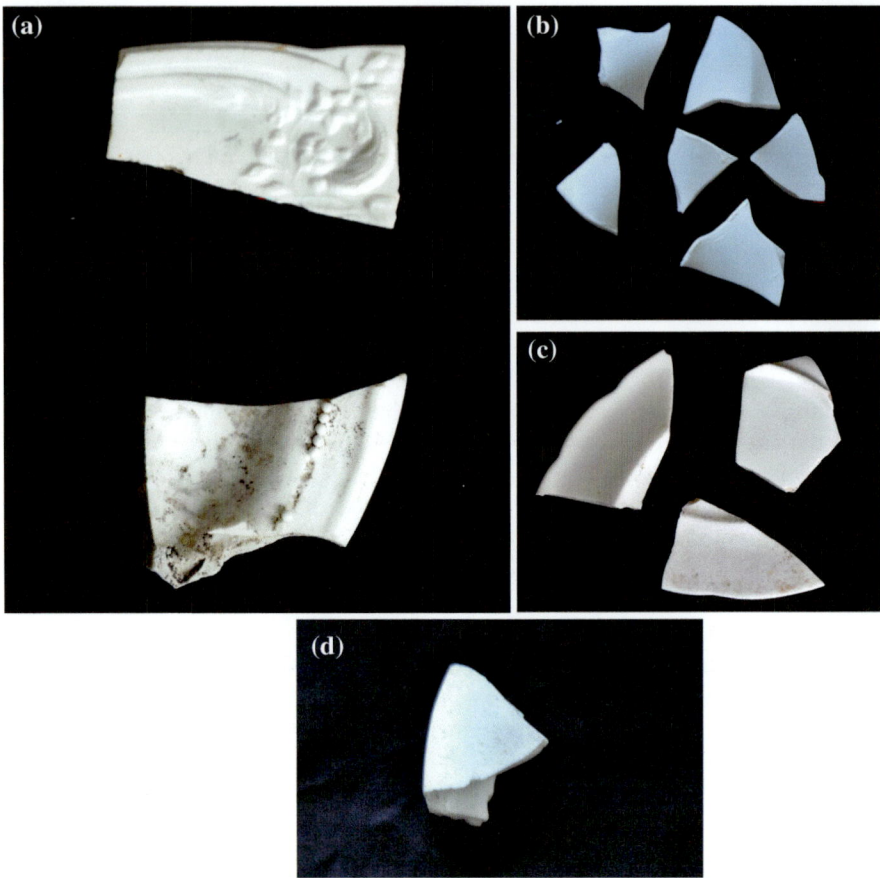

Fig. 4.3 a–d Nantgarw porcelain shards from the excavation of the Nantgarw China Works site in 1995 showing typical mouldings (**a**) which can be matched with completed and perfect Nantgarw china specimens still extant. The variety of sizes and shapes (**b** and **c**) can be seen in selected shards, and the effect of overfiring in the biscuit kiln which has produced a sagging and warping of the porcelain in the resultant piece can clearly be seen in shard (**d**). Reproduced with permission of the Nantgarw China Works Trust, Nantgarw China Works Museum, Tyla-Gwyn, Nantgarw

An invoice in the possession of the Nantgarw China Works Museum, dated 1855, describes William Henry Pardoe as a "*manufacturer of breakfast, dinner, dessert and tea porcelain services, decorated with arms, crests, cyphers etc. and other decoration, along with clay pipes, garden and flower pots, vases, pipes for conveying water and other articles of pottery ware*".

However, it seems that this re-birth of the porcelain enterprise at Nantgarw was once again to be short lived. Firstly, the *Bristol Mercury* issue of the 16th August 1856, announces an advertisement for the appointment of a skilled potter and also the disposal of equipment used in the Pottery:

Potters: Wanted a man to dip and set a few kilns of biscuit ware now on a factory about to be closed. For sale: three shafts for pug mills, three foot wheels for red ware; pantile, brick and cress moulds; good chaff cutting machines; strong cart; lot of dryers for pipemakers; garden pots worth the attention of nurserymen; new jigger and rotary lifting machine; firebricks; bevel wheel and pinion; lot of best and second ware, with several other articles used in pottery business, all of which will be sold cheap, a clearance.

Apply to William Henry Pardoe, Nantgarw Pottery, Near Cardiff

Secondly, the *Usk Observer* of May 15th, 1858, gives notice that William Henry Pardoe was selling off his remaining stock of porcelain at Nantgarw including *"elegant tea services decorated and in the white"*—a repetition of the events which had occurred at the Nantgarw China Works some 37 years previously! After this date there is no further mention of the manufacture of porcelain at Nantgarw. The first announcement of 1856 in the *Bristol Mercury* is interesting on two counts as, firstly, it describes the presence of a few kilns' worth of biscuit china ware that needed to be glazed prior to sale—confirming that some kind of porcelain had been made there in reasonable quantity (a typical kiln load seemed to be around 25 dozen plates and items of tea and dinner ware). Also, the advertisement offers the sale of firebricks at Nantgarw—some thirty-five years after the initial trials of the original silica firebrick had been initiated there by William Weston Young. So, this could imply strongly that either Pardoe was selling off Young's experimental firebrick stock from the 1821/22 period (which surely would not have been significant) or, alternatively, that he had himself embarked upon the production of silica firebricks. This has not been alluded to hitherto and we have no historical documentation to offer a lead—but it could mean that firebricks may have been made at Nantgarw long after William Weston Young had departed for his successful commercial Vale of Neath refractory silica brick enterprise.

An archaeological survey of the Nantgarw China Works site in a 1931 survey (Williams, *The Nantgarw Pottery and its Products: An Examination of the Site* 1932) concluded that Kiln I on site was a porcelain biscuit kiln dating from the Billingsley and Walker era and Kiln II was a porcelain glost kiln, which seemed a reasonable conclusion at the time. However, a more recent survey conducted in 1995 (Murphy et al. 1997) determined that Kiln II could not have been from Billingsley's era as later clay pipes and earthenware shards manufactured in William Henry Pardoe's era were found embedded in the structure—it seems quite reasonable to conclude therefore that this kiln was erected by Pardoe for his later venture into porcelains. Hitherto, it had been concluded by Williams that a third kiln, Kiln III, was erected during Pardoe's time as it was of an inferior construction and state of preservation which would not have been considered compatible with Samuel Walker's exacting kiln standards. It is possible of course that this Kiln III could have been the kiln adopted by William Weston Young for the firing experiments on his silica bricks which he undertook at Nantgarw—in fact, had this been the original glost kiln of Billingsley and Walker and adapted by William Weston Young to take the higher firing temperatures of his silica brick specimen ceramic trials, the additional work undertaken at elevated temperature for the experimental firebricks may have been sufficient to cause more stress damage to the existing brick linings which had not been designed for this work, hence resulting in the observed increased damage to this kiln.

The intriguing question now, of course, is what chemical composition Pardoe used for his porcelain recipe in the 1850s? He may have had access to his father's notes on the experiments of William Weston Young, carried out between 1820 and 1821 in the presence of Pardoe at Nantgarw, or indeed he may have had access to Young and Wright's business venture into the *"Nantgarw dry mix"* creation of 1835 which proved to be an unpopular commercial venture at that time: indeed, he could perhaps have even purchased some batches of this *"Nantgarw dry mix"* for his own evaluation at Nantgarw. Either way, Pardoe's recipe would unlikely to have been based upon the true Nantgarw soft paste variant as this would have had equally bad production problems for him in the 1850s even with the potential then available for much better kiln temperature control.

William Henry Pardoe died in 1867, leaving a widow and several sons; his widow, Mary, described as a *"potteress"* ran the Nantgarw factory after the death of her husband, employing 11 men and 10 women at Nantgarw, and trading as Pardoe Bros. (namely, Percival and Felix Pardoe, born in 1851 and 1852, respectively) between 1871 and 1895. Then, Percival Pardoe ran the declining factory until his death in 1920. His descendants continued to live in Nantgarw House until the 1970s (where William Billingsley originally lived with his daughter Lavinia and the widowed Samuel Walker); this fell into disrepair and has now been restored as the home of the Nantgarw China Works Museum Trust who are resourcing the conservation and preservation of the site for posterity.

References

H.G.M. Edwards, *Swansea and Nantgarw Porcelain: A Scientific Reappraisal* (Springer, Dordrecht, 2017)

I.S. Evans, K. Jones, *Quantitative Geography: A British View* (Routledge & Kegan Paul, Boston, 1981), pp. 123–134

K. Murphy, R. Ramsey, D.A. Higgins, The dismantling of Kiln II, Nantgarw China, Pottery and Pipe Works, Mid-Glamorgan, 1995. Post Mediaeval Archaeol. **31**(1), 231–247 (1997)

I.J. Williams, *The Nantgarw Pottery and its Products: An Examination of the Site* (The National Museum of Wales and the Press Board of the University of Wales, Cardiff, 1932)

R. Williams, *Nantgarw Porcelain 1813–1822: Inaugural Lecture Given to the Friends of Nantgarw China Works Museum* (GPS Printers, Taff's Well, Pontypridd, 1993)

Chapter 5
Samuel Walker

Abstract The interaction between William Weston Young and Samuel Walker from their earliest days at the Nantgarw China Works through the Memorial of September 1814 and finally to the phase II of the re-establishment of porcelain production at Nantgarw until the departure of Walker first for Coalport in 1820 and thereafter for the USA.

Keywords Samuel Walker · Nantgarw China Works · Memorial of September 1814 · Coalport China Works · Temperance Hill Pottery · Nantgarw paste formulation

The Samuel Walker of our study shares his name with the eponymous and perhaps better known, Captain Samuel Walker (1817–1847) of the US Army Mounted Rifles, who served in the American Indian Wars and in the Mexican-American War; a literature search on "Samuel Walker" invariably comes up with references to the military man rather than the ceramics kiln engineer. The military Samuel Walker was captured during the Mexican invasion of Texas led by Santa Anna and he escaped to join the Texas Rangers in 1844, where he was under the command of a Captain Jesse Billingsley (!!)—despite the unusual name, this man does not seem to have any connection with our William Billingsley. Walker was killed in action leading his men during an attack on Huamantla in Mexico and was later re-interred at San Antonio, Texas, after the cessation of hostilities. He is famous for his collaboration with Samuel Colt, the arms manufacturer, in the production of the Walker Colt 0.44 (11.5 mm bore) calibre revolving pistol in 1847, a six-shot ball and percussion cap weapon which was based on the earlier Paterson Colt with improvements in trigger guard, rapid reloading capability, better sights, a robust mainframe and loading retractor lever mechanism. For a while this weapon, illustrated in Fig. 5.1, of which some 1100 only were made for the Texas Rangers, which weighed in at a hefty 2 kg unloaded, was the best revolver created until it was later superseded by the Colt 1851 Navy model in a smaller 0.36 calibre, which made it lighter (approximately half the weight) and more tactile to handle. The Walker Colt was described as the "*world's right arm*" and was far and above much better than anything else on the market at that time—it was seemingly therefore the equivalent in its genre to the standing of Nantgarw porcelain in the ceramics field and, like the best Nantgarw porcelain, its

© Springer Nature Switzerland AG 2019

H. G. M. Edwards, *Porcelain to Silica Bricks*,

https://doi.org/10.1007/978-3-030-10573-0_5

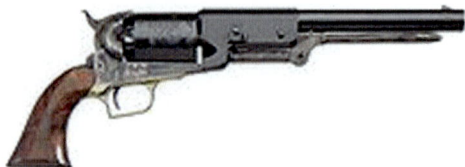

Fig. 5.1 A Walker Colt revolving pistol, 0.45 calibre, designed by the eponymous Samuel Walker for the Texas Rangers in 1847. In its genre this was the equivalent of the finest Nantgarw porcelain, as when it was created it had no match from any of its competitors. The pistol was replaced by Samuel Colt in the Colt Model 1851 a few years later, in a lighter, smaller 0.36 calibre and more easily handled version

rarity and desirability make it a very expensive antique to own today—both of these typical specimens are associated with the name of "Samuel Walker", although, of course, they are different people!

The life of our Samuel Walker has already been described elsewhere in this book, demonstrating his close involvement with William Billingsley, his ceramics mentor and father-in-law. From the time of their first meeting at Brampton-in-Torksey in 1806, where his family and that of William Billingsley were residents on neighbouring farms, the association between the two men grew in stature: Walker, an engineer, had a flair for the design and construction of ceramic kilns which must have been a godsend for William Billingsley in his efforts to create the finest porcelains. In 1812, whilst engaged at the Royal Worcester China Works. Walker was commissioned by Martin Barr personally to design and build a new reverberatory kiln for the manufacture of porcelain at high temperatures: the work was deemed to be so secret that it had to be accomplished at night and behind closed doors. This was achieved alongside Barr, Flight and Barr, through the offices of Martin Barr, in gratitude granted Walker and Billingsley the sum of £200 in 1813 with the stipulation that "they forever hereafter forebear from communicating and imparting the secret relating to a new method of composing porcelain" (John, *William Billingsley* 1968, p. 36; Nance, *The Pottery and Porcelain of Swansea and Nantgarw* 1942, p. 238). A hidden secret of ceramics manufacture was the maintenance of a steady high temperature for several days in the kiln, with only a small temperature gradient being applied, and kiln design was critical for the firing of batches of porcelain which frequently would comprise many dozens, if not hundreds, of items placed on racks and separated by ceramic saggars or spacers. William Billingsley has commented in a letter to John Coke, his china works proprietor at Pinxton, around 1800 that some 26 dozen plates would be fired in one batch in the porcelain biscuit kiln. The loss that would occur in getting the kiln temperature incorrect or the firing details wrong would therefore be immense, so the kiln manager was a respected and valued member of the production team (Edwards, *Swansea and Nantgarw Porcelain: A Scientific Reappraisal* 2017).

Martin Barr died on November 10th, 1813, and Billingsley and Walker thereupon lost his personal support at Worcester: Barr's partners at Worcester, Messrs Flight and Barr, having now also taken on George Barr to become Flight, Barr and Barr, were not interested in exploring novel methods of porcelain production, as was Martin Barr, and the committing of investment capital and Flight, Barr & Barr wrote to Walker that Billingsley and Walker's secret porcelain formula *"could not profit any partners"* (Nance, *The Pottery and Porcelain of Swansea and Nantgarw* 1942, p. 240)—this surely was the catalyst for the immediate departure of Billingsley and Walker from Worcester in late 1813/ early 1814. After the collapse of the initial Nantgarw China Works enterprise in late 1814 after the unsuccessful submission of the Memorial to the British government, announced by letter to Sir John Nicholl of Merthyr Mawr on September 12th, 1814, Walker and Billingsley were engaged at the Swansea Pottery by Lewis Weston Dillwyn. Walker was appointed the sole china maker at the fledgling Swansea China Works, still called the Swansea Pottery, in October 1814 and he was put in charge of new porcelain manufacture, whilst William Billingsley was put in charge of the decorating shop. It is understood that Billingsley still continued his experiments with his novel porcelain but Dillwyn is recorded as saying that he doubted they would ever be successful. This could account for the statement that has appeared in several historical accounts that Dillwyn *"manufactured Nantgarw porcelain at Swansea"* and for the generation of the threatening letter from Flight, Barr and Barr (and not from Martin Barr as is often incorrectly quoted), warning Dillwyn against the employment of Billingsley and Walker to make the novel porcelain there upon the forfeit of a penalty of £1000! Note that this letter appeared in late 1814 from the Royal Worcester China Works and could not have been initiated by Martin Barr, who had died a year previously—contrasting with several historical attributions otherwise.

Billingsley, evidently tired of the limitations imposed upon him by Dillwyn and possibly the direction being taken at the Swansea Chinas Works, eventually left Dillwyn's employment at Swansea on December 23rd, 1816, leaving Walker there alone to manage Dillwyn's kilns for the firing of the *trident* body: disappointed in Dillwyn's retrograde step in refusing to make the esteemed duck-egg body any more at Swansea, Walker wrote of his dissatisfaction to Josiah Wedgwood, seeking an opportunity at his Etruria works in Stoke-on-Trent. None was available, and Walker departed the Swansea China Works eventually on September 28th, 1817, to join Billingsley who had started to set up again to manufacture porcelain in Nantgarw. A double tragedy struck both families that year in that Sarah Walker died in Swansea on January 1st, 1817, and her younger sister Lavinia Billingsley died in Nantgarw on September 18th, 1817, aged just 20, just ten days before Walker arrived there from Swansea. The second phase operation at Nantgarw was always small and Richard Millward, who was employed as a child at the Nantgarw China Works, revealed to William Turner for his book on the Nantgarw factory published in 1897 (*The Ceramics of Swansea and Nantgarw* 1897), that in 1819 at the height of its success only 8 adults and 12 children were employed there.

Billingsley and Walker stayed together through severe trials and much tribulation until Billingsley died in Coalport in 1828. After that, the documentary history of Samuel Walker is rather sketchy until he decides to emigrate to the USA in the late 1840 s. However, it appears that he did try to revive the production of porcelain by Lewis Weston Dillwyn at Swansea in the late 1820s and had approached Dillwyn with this idea but Dillwyn had lost interest in such a venture at this stage and had decided to concentrate instead on a local and national political career. A rather shadowy figure in the Swansea China Works is George Haynes, who apparently was very much an autocratic manager, appointed by Lewis Dillwyn Sr. to oversee the factory and look after the young Lewis Weston Dillwyn, who nominally at least was the factory proprietor. Haynes was closely involved in an advisory capacity to William Weston Young, as recorded in Young's diaries, on the prospect of Young being made a partner in the Phase I Nantgarw operation along with Walker and Billingsley seemingly outside the knowledge of Dillwyn.

The life story of Samuel Walker now becomes a mystery for some years after 1828, until he re-surfaces again in 1842: it can be presumed that perhaps he was still engaged somewhere in porcelain manufacture, in Coalport for an indeterminate time and then possibly in Staffordshire, as in 1847 John Taylor, a Staffordshire potter, published a book entitled "*The Complete Potter*", whose title hides a very important piece of information relating to the Nantgarw China Works: Taylor provides the first written evidence of Billingsley and Walker's Nantgarw porcelain body recipe, which had remained such a closely guarded secret for over twenty-five years, which had been given to him personally by Samuel Walker. It is interesting that Walker did not attempt to sell this recipe or to have made porcelain from it in the intervening years, and one can merely suggest that the commercial non-viability of the formulation with its unacceptably high kiln wastage upon firing still preyed heavily upon him. Alternatively, he could still have been concerned at the written threat posed by Flight, Barr and Barr to exact a financial penalty if he embarked upon the production of Nantgarw porcelain with a third party for financial gain?

Samuel Walker, then aged about 50, with his second wife Sarah (aged 27) and their daughter Mariah (aged 8), emigrated from Liverpool on 22nd April 1842, aboard the ship *Monument,* and arrived in New York a month later. He established the Temperance Hill Pottery in Troy, New York State, manufacturing teapots, pitchers and toys in the so-called "*Rockingham*" brown salt-glazed earthenware (Broderick, *An English Porcelain Maker in West Troy* 1988). He leased a house in West Troy on June 26th, 1849, at $6 a year for the first five years and then for $10 a year thereafter, finally giving up the property in 1869. The house still exists and recent excavation in the garden has unearthed several shards of glazed white porcelain, complete pieces of china and "*Rockingham*" type earthenware which conclusively indicates that porcelain was being made there. In 1860–62, Walker moved to Norwich, Connecticut, where he worked at a local pottery until his wife died on September 16th, 1862, after which Walker went to live with Mariah and her family, until he died in 1875 aged 80.

References

W.F. Broderick, An English porcelain maker in West Troy. Hudson Valley Reg. Rev. **5**(2), 23 (1988)

H.G.M. Edwards, *Swansea and Nantgarw Porcelain: A Scientific Reappraisal* (Springer, Dordrecht, 2017)

W.D. John, *William Billingsley* (Ceramic Book Co., Newport, 1968)

E.M. Nance, *The Pottery and Porcelain of Swansea and Nantgarw* (Batsford, London, 1942)

J. Taylor, *The Complete Practical Potter* (Shelton, Stoke-upon-Trent, 1847)

W. Turner, *The Ceramics of Swansea and Nantgarw* (Bemrose & Sons, Old Bailey, London, 1897)

Chapter 6
Lewis Weston Dillwyn

Abstract The role of Lewis Dillwyn in shaping the early career of William Weston Young and his later influence on Young through the Swansea China Works and especially his interactions with William Billingsley and Samuel Walker, and thereby impacting upon Young at the Nantgarw China Works.

Keywords Dillwyn · Swansea China Works · Young · Pardoe · Billingsley · Walker · Duck-egg porcelain · Trident porcelain · Alfred Russel Wallace

Lewis Weston Dillwyn was born on August 6th, 1778, in St. John, Hackney, London, the eldest son of William Dillwyn and his cousin Sarah Weston, who were married in 1777. William Dillwyn was born in Delaware County, Pennsylvania, in 1743 and he came to England in 1777 for the express purpose of marrying Sarah: being a prominent member of the Society of Friends in Philadelphia and hence an avowed non-combatant, William received special dispensation from General George Washington to pass through American lines and he was able to depart New York freely. On July 6th, 1777, he arrived in Cork and then took a sailing packet to Swansea, where he first visited the Cambrian Pottery before proceeding on to Bristol to marry Sarah. Swansea seemed to hold a fascination for him, particularly the Cambrian Pottery, where he stayed for several months in 1801 and 1802; it has been suggested with a view to possibly purchasing the Pottery from George Haynes. Dillwyn Sr. believed that the Swansea Pottery had great potential for expansion and for increasing the sales of its wares through adoption of an agent, which had not occurred hitherto as they had sold their product locally and directly from the factory. In 1797, Lewis Dillwyn was apprenticed to a haberdasher in Dover, but William believed that the opportunity offered by his purchase of the Cambrian Pottery was much better for him and in 1802 father and son acquired the Swansea Pottery, as it came to be known. An added incentive for Lewis Dillwyn was the rich natural history that abounded in the coastal region and the hinterland around Swansea: the son seemed to have inherited his father's love of botany in particular and the relatively unexplored areas around Swansea at that time were a great attraction for them. However, because Lewis Dillwyn was still young, then aged 24 years, and had no experience of managing a pottery, so George Haynes at 57 years old was enlisted for seven years to provide the business acumen and wherewithal to run the pottery alongside Dillwyn Jr., as at this

© Springer Nature Switzerland AG 2019

H. G. M. Edwards, *Porcelain to Silica Bricks*,

https://doi.org/10.1007/978-3-030-10573-0_6

time William Dillwyn had left Swansea and had returned to his family seat at High Hall, Essex. A large amount of money had been paid by Dillwyn Sr. for the purchase of the Swansea Pottery, with claims of between £12,000 and £16,000 being cited. It seems clear that George Haynes actually still managed the Swansea Pottery as Lewis Dillwyn was much taken up with his natural history studies and between 1802 and 1809 he devoted most of his attention to preparing his book on *The British Confervae*, which saw him elected to the Royal Society and the Linnaean Society in 1804. It was in this book that William Weston Young contributed two much-esteemed drawings, the first being that of *Conferva dissiliens* (plate 63) from Crymlyn Bog, Swansea, and the second that of *Conferva youngana* (plate 102), which bore his name, from limestone rock pools at Dunraven Castle. At the end of the seven years appointment of George Haynes, Dillwyn was deemed still not ready to take over the management of the Swansea Pottery, so Haynes' appointment was extended further until 1810. We have seen that he was still there in July 1814 when he was asked by Young to deliver his opinion on the viability of the Nantgarw China Works, of whose existence he seemingly never sought to apprise Lewis Dillwyn.

During this time, Dillwyn decided that the style of flower painting at the Swansea Pottery needed to be changed to reflect a more lifelike representation rather that the stiff, stylised rendition that had been undertaken there: the principal painter of botanical subjects at the Swansea Pottery from 1795 had been Thomas Pardoe and under the influence of Lewis Dillwyn, Pardoe's painting style changed to become much more realistic. It is thought that Pardoe stayed on at Swansea until late 1811 or perhaps even 1812, whence thereafter he was based at Bristol working up his own ceramics decorating business until he was tempted away by William Weston Young to work on decorating the Nantgarw remaining stock. Of course, during this time also we have noted the appearance at the Swansea Pottery of William Weston Young, fresh from the failure of his agricultural supplies business and newly bankrupt, who was engaged by Dillwyn to paint botanical subjects on his earthenware. The initial magnetism between Young and Dillwyn was no doubt generated by their shared love of natural history, of flora and fauna, and Young's contribution to Dillwyn's chef d'oeuvre *The British Confervae* (1804–1809) has been illustrated earlier. Nevertheless, Young's painting alongside that of Pardoe can be recognised at the Swansea Pottery—and occasionally, their efforts are to be found together on the same piece.

The next part of the story concerns Dillwyn's attempts to make porcelain at Swansea (Dillwyn, *Notebooks* 1922 and cited in Eccles and Rackham, *Analysed Specimens of English Porcelain* 1922; Appendix B): initially, there is reference in the literature (Chaffers, *Marks and Monograms on English Porcelain with Historical Notes on each Manufactory* 1863; Jewitt, *The Ceramic Art in Great Britain* 1878; Turner, *The Ceramics of Swansea and Nantgarw* 1897) to a curious "*opaque china*" which was produced there around 1810. This could be related to the use by George Haynes of soaprock in the earthenware recipe, the effect of which he much admired, for which Lewis Dillwyn had to go to Cornwall personally to acquire supplies directly from the mine (Nance, *The Pottery and Porcelain of Swansea and Nantgarw* 1942). Whereas this assertion is of course conjectural, the addition of soaprock or steatite, a magnesium silicate, formulated as $MgSiO_3$, to the clay for firing earthenwares would be expected to give perhaps a glassier and denser finish to the pot, but problems in adherence of the glaze were also noted. It is fascinating to recall that just some seven years or so later, Lewis Weston Dillwyn would again revert to the use of soaprock as an additive component in his trial experiments to produce a new paste in an unsuccessful attempt to market a more robust version of his esteemed duck-egg paste—giving the Swansea *trident* ware, much of which has been written about before (Edwards, *Swansea and Nantgarw Porcelains: A Scientific Reappraisal* 2017; John, Sw*ansea Porcelain* 1958; Jones and Joseph, *Swansea Porcelain* 1988).

Dr. Maurice Hillis in his authoritative thesis on the development of Welsh porcelain bodies (Hillis, *Welsh Ceramics in Context* 2005, Vol. II, Chap. 9) has researched the relationship between the chemical analyses of porcelain shards recovered on or near factory sites in Swansea and Nantgarw and has compared these with surviving recipes. He has made an intriguing discovery about the visit of a Dr. Donovan to Swansea in 1801/2, who commented that *"various kinds of earthenware and porcelain are manufactured there"* some twelve years before Dillwyn is recorded as establishing the Swansea China Works for that express purpose. Donovan goes on to say that soapstone was pulverised, moistened and mixed with a fine clay to produce an *"elegant porcelain such as that made at Bristol, Vauxhall, Worcester, Liverpool and Caughley"*. Hillis suggests that the porcelain referred to by Donovan was probably the glassy pearlware, or something called *"stone porcelain"*, then being made in Swansea, which could have mistakenly have been thought to have been a porcelain body to a casual observer, but that George Haynes then must have had the idea of using soaprock in his mixture as he soon despatched Lewis Dillwyn to Cornwall to obtain some. In a statement made by Henry Morris, who was an apprentice at Swansea later in the decade, to Colonel Grant-Francis in 1850 (later published in the *Cambrian newspaper* on 3rd January, 1896), he noted that two ex-Coalport China Works employees, named Burn and Briggs, were then taken on by Dillwyn in 1811/12 to assist with the proposed manufacture of porcelain at Swansea—at which they proved unsuccessful and were promptly dismissed. So, clearly, Dillwyn was in the market for hiring porcelain experts at that time to kick-start his fledgling Swansea China Works project and, as we know, was alerted to the presence of Billingsley and Walker, who were then based at their troubled Nantgarw China Works, by Sir Joseph Banks in September 1813. It is equally clear that no form of true or hybrid porcelain was made at Swansea before the arrival of Billingsley and Walker. Hillis also considers the possibility that Billingsley and Walker actually brought with them to Swansea their formulaic recipe for Nantgarw porcelain which was given to Dillwyn: this is very conjectural, bearing in mind the strict secrecy shrouding the Nantgarw porcelain recipe and the threat posed by Flight, Barr and Barr to a penalty imposed upon them if they released the details of this recipe to a third party. We shall return to this possibility later when the analytical data for Nantgarw shards are examined and interpreted and the composition of the early Nantgarw recipe is evaluated.

Nance (*Pottery and Porcelain of Swansea and Nantgarw* 1942) has written extensively about the interaction between Young and Pardoe at the Swansea Pottery—both were very fine painters indeed, and clearly their friendship, formed in these years between 1802 and 1806 when both painted at Swansea under Dillwyn's tutelage, stood Young in good stead for the later employment of Pardoe in his attempt to recover the fortunes of the Nantgarw China Works in 1820–1823. Young and Pardoe parted company with Lewis Dillwyn in 1806 and 1811/12, respectively, but must have stayed in contact to enable Young to quickly seize control of the stock of the Nantgarw China Works in 1820 after the departure of Billingsley and Walker for Coalport and to commence its decoration by Thomas Pardoe for the auction sales. We have noted the frequency with which Pardoe's name appears in the Young Diaries and to the regular and sometimes quite significant monetary payments made by Young to Pardoe in Bristol for china decoration.

6.1 Alfred Russel Wallace: A Link Persona

Born on the 8th January 1823, in Llanbadoc, Usk, and died at the age of 90 in 1913, Alfred Russel Wallace may at first sight have no connection with the main players in our theme centred on William Weston Young. The chronology seems wrong and for many years he travelled overseas in such far-flung places as the Amazon and the East Indies. There is no documentation that he ever met with William Weston Young, yet they shared several commonalities which indicate that they were both of a similar restless disposition and with their location in Neath and engaged on similar activities is a remarkable coincidence. Some features they both shared (Raby, *Alfred Russel Wallace: A Life* 2001) are as follows:

Both were rather reserved and shy in public, reticent in their own lives, but determined to see through projects to completion: setting standards that were tough on themselves.

They both frittered away money in disastrous financial deals.

They both loved natural history, flora, fauna and particularly botany: with especial regard for the wildness and beauties of the Vale of Neath. Both elected members of the Linnaean Society for their botanical prowess. Wallace subscribed to Darwin's statement in his book, *The Voyage of the Beagle*, published in 1835: "*Hence, a traveller should be a Botanist for in all views plants form the chief embellishment*" (later included in Darwin and Fitzroy 1839).

Both were land surveyors based in Neath: Young in Fairyland and Wallace in Llantwit, where both rented properties with substantial gardens and outbuildings overlooking the River Nedd and the Neath Canal. In Wallace's time there were already eight land surveyors in Neath in a town of 6000 inhabitants, but he managed to acquire several good contracts despite the competition.

Both attended at the Society of Friends Meeting House but neither had a deep-rooted interest in formal religion: in contrast with Wallace's friend Charles Darwin, who was tormented by the perceived divide between evolutionary science and religious beliefs.

Both surveyed the Vale of Neath for the establishment of a railway: in Wallace's case, to bring coal and iron to the port of Swansea from Merthyr Tydfil and in Young's to effect transportation of his own firebricks to Swansea and its docks.

In the 1840s, Wallace was engaged with the Mackworths of the Gnoll Estate to found a Mechanics Institute in Church Place, Neath, opening in 1848 and which is now the Neath Museum. As Curator of the Philosophical and Scientific Institute in Neath, Alfred Wallace came into contact with Lewis Dillwyn of Sketty Hall, Swansea, with whom he shared a love of botany and natural history. When the British Association came to Swansea in 1848, Dillwyn gave Wallace an honourable mention in a paper on the flora and fauna of Swansea he presented at the meeting. The same year, Wallace embarked the *Mischief* at Liverpool on the 26th April 1848, and sailed for South America, arriving in Salinas, Brazil at the mouth of the Amazon on the 26th May 1848, where he stayed for some years researching new species of plants and insects. In 1858, Wallace, aged 35 and suffering from malaria in the East Indies, wrote to Charles Darwin, saying that he had worked out a theory of natural selection. Darwin, who had been secretly working on this issue for decades, was aghast: in just two weeks, Darwin and Wallace's papers were presented jointly at the Linnaean Society in London and in 1859, Darwin published his famous "*On the Origin of Species by Natural Selection*", which set the foundation for evolutionary theory and brought him into conflict with church leaders, particularly Samuel Wilberforce (christened "Soapy Sam" by T. H. Huxley), the Bishop of Oxford. Most modern historians regard Wallace as a neglected giant of the history of science and ideas and his work on the plurality of worlds (Wallace, *Man's Place in the Universe: A Study of the Results of Scientific Research in Relation to the Unity or Plurality of Worlds* 1904) is still regarded as a classic text which fully endorses his intellectual prowess in addressing the bigger philosophical questions.

It is appreciated that William Weston Young and Alfred Russel Wallace would not have met each other as the chronology and their respective geographic locations would have dictated otherwise, nevertheless a distinct similarity exists between their interests and it is quite probable that Dillwyn would have discussed with Wallace his prior interaction with Young, particularly with regard to their shared love of local natural history, specifically of Swansea and the Vale of Neath.

References

W.B. Chaffers, *Marks and Monograms on Pottery and Porcelain with Historical Notes on Each Manufactory,* 1863, *J. Davy & Sons, London,* Kessinger Legacy Reprints (Kessinger Publishing, Whitefish, 2010)

C. Darwin, *On the Origin of Species by Natural Selection or The Preservation of Favoured Races in the Struggle for Life* (John Murray, Albemarle St., London, 1859)

C. Darwin, R. FitzRoy, *Narrative of the Surveying Voyages of His Majesty's Ships Adventure and Beagle Between the Years 1826 and 1836 Describing their Examination of the Southern Shores of South America and the Beagle's Circumnavigation of the Globe,* 4 vols., Henry Colburn, London, 1839. (Darwin's "*The Voyage of the Beagle*" comprises Volume 3 of this set: it became so popular that it was re-issued as a separate volume in its own right and as a 2-volume issue in 1846: for example C. Darwin, R. Fitzroy, *Journal of Researches into The Natural History and Geology of the Countries Visited During the Voyage of HMS Beagle Round the World Under the Command of Catain Fitzroy, RN*, 1st edn., 2 vols., Harper & Bros., New York, 1846)

L.W. Dillwyn, *British Conferva or Coloured Figures and Description of British Plants Referred by Botanists to the Genus Conferva* (W. Phillips. George Yard, Lombard St., London, 1809)

L.W. Dillwyn, *Notes on the Experimental Production of Swansea Porcelain Bodies and Glazes.* Made by Lewis Weston Dillwyn with Samuel Walker at the Swansea China Works Between 1815 and 1817. Presented to the Library of the Victoria &Albert Museum, South Kensington, London by John Campbell in 1920. Reproduced in Eccles & Rackham, *Analysed Specimens of English Porcelain, 1922*, see reference below

H. Eccles, B. Rackham, *Analysed Specimens of English Porcelain in the Victoria and Albert Museum* (Victoria and Albert Museum, London, 1922)

H.G.M. Edwards, *Swansea and Nantgarw Porcelain: A Scientific Reappraisal* (Springer, Dordrecht, 2017)

M. Hillis, The development of Welsh porcelain bodies, in *Welsh Ceramics in Context*, vol. II, ed. by J. Gray (Royal Institution of South Wales, Swansea, 2005), pp. 170–192

L. Jewitt, *Ceramic Art in Great Britain* (Virtue & Co., London, 1878)

W.D. John, *Swansea Porcelain* (Ceramic Book Co., Newport, 1958)

A.E. Jones, Sir L. Joseph, *Swansea Porcelain: Shapes and Decoration* (D. Brown and Sons, Ltd., Cowbridge, 1988)

E.M. Nance, *The Pottery and Porcelain of Swansea and Nantgarw* (Batsford, London, 1942)

P. Raby, *Alfred Russel Wallace: A Life* (Chatto & Windus, London, 2001)

W. Turner, *The Ceramics of Swansea and Nantgarw* (Bemrose & Sons, Old Bailey, London, 1897)

A.R. Wallace, *Man's Place in the Universe: A Study of the Results of Scientific Research in Relation to the Unity or Plurality of Worlds* (McClure, Phillips & Co., New York, 1904)

Chapter 7
William Billingsley

Abstract The influence of William Billingsley and his objective of creating the world's finest porcelain at the Nantgarw China Works on William Weston Young: the support of Young for Billingsley at Nantgarw and the actions of Young following Billingsley's departure for the Coalport China Works in 1820.

Keywords William Billingsley · Nantgarw China Works · Swansea China Works · The Memorial of September 1814 · Kiln wastage · Porcelain decoration

Much has already been written about William Billingsley, whose vision it was to synthesise the finest porcelain the world had ever seen at Nantgarw; dedicated texts include those of John (*William Billingsley* 1968), Edwards and Denyer (*William Billingsley: The Enigmatic Porcelain Artist, Decorator and Manufacturer* 2016), and Robinson and Thomas (*Not Just a Bed of Roses: The Life and Work of the Artist, Ceramicist and Manufacturer, William Billingsley, 1758–1828* 1996) whilst other works devote major sections to a description of his work at several porcelain factories, such as John (*Nantgarw Porcelain* 1948, *Swansea Porcelain* 1958), Jones and Joseph (*Swansea Porcelain* 1988), Edwards (*Nantgarw Porcelain: The Pursuit of Perfection* 2017a, *Swansea Porcelain: The Duck-Egg Vision of Lewis Dillwyn* 2017b, *Swansea and Nantgarw Porcelain: A Scientific Reappraisal* 2017c, *Derby Porcelain: The Golden Years: 1770–1830* 2017d, *Nantgarw and Swansea Porcelain: An Analytical Perspective* 2018), Twitchett (*Derby Porcelain 1748–1848* 2002), Gent (*Pinxton Porcelain* 1996), Nance (The *Pottery and Porcelain of Swansea and Nantgarw* 1942), Turner (*The Ceramics of Swansea and Nantgarw* 1897) and Haslem (*The Old Derby China Factory* 1876). In the light of this considerable available literature, therefore, only the salient features relating to the interaction between William Billingsley and William Weston Young will be addressed here.

William Billingsley was born in 1758, in the parish of St Alkmunds, Derby, as was Thomas Pardoe, and was apprenticed to the Derby China Works under William Duesbury in 1775, the year before William Weston Young was born: Billingsley was therefore some 18 years older than Young. William Billingsley would have first met Thomas Pardoe at the Derby China Works, where Billingsley was head of the enamelling workshop in which Pardoe was apprenticed. In 1795, Billingsley left Derby to assist John Coke in the manufacture of porcelain at Pinxton in Derbyshire, but

© Springer Nature Switzerland AG 2019
H. G. M. Edwards, *Porcelain to Silica Bricks*,
https://doi.org/10.1007/978-3-030-10573-0_7

after a few years he left Pinxton to set up on his own at Mansfield and thereafter at Brampton-in-Torksey, where he decorated porcelain bought in from other factories, including Derby and Coalport. Here, Samuel Walker, who lived in a neighbouring farm, courted and eventually married Billingsley's eldest daughter, Sarah, in Worcester, where they took up employment with Martin Barr of Barr, Flight and Barr in 1808. At Brampton-in-Torksey, Walker and Billingsley developed ideas and experimented with porcelain manufacture and kiln design, which they fully exploited in their time at Worcester with the full cognisance and encouragement of Martin Barr; it seems that they made small amounts of a porcelain of their own composition at Worcester and modified the kiln processes accordingly. They departed Worcester to set up on their own at Nantgarw, parting company with Martin Barr in amicable circumstances: he even bestowed upon them a gift of £200 to assist in the manufacture of their own porcelain. Some authors have suggested that the catalyst for their departure from Worcester was the death of Martin Barr, who had strongly supported their efforts to create a better quality porcelain than was currently being made there: in contrast, Messrs. Flight, Barr and Barr, who had assumed control of the Royal Worcester China Works upon the death of Martin Barr, apparently had no interest at all in incurring the expense of venturing into novel porcelain production and wished to concentrate their efforts on the sale of their own product in the market place.

At this point we need to investigate when William Billingsley and Samuel Walker actually met William Weston Young since many historical accounts actually ascribe Billingsley and Walker's selection of the Nantgarw site to Young's advice—and he would have known of the availability and advantages of this particular site for porcelain manufacture from his land surveying in Glamorgan, a position which he had held since 1811. It is believed that Billingsley and Walker first arrived in Nantgarw in January 1814, when formally Young has recorded their meeting in January—but did they meet Young earlier when possibly he would have been able to apprise them of the availability and advantages of the Nantgarw site? The *Memorial*, signed by Billingsley (as his diminutive "Beeley"), Walker and Young, in which they sought financial support to the extent of £500 from the British Government for their Nantgarw Phase I enterprise is dated September 5th, 1814; it is also recorded in his Diaries that Young first met with Billingsley and Walker at the Nantgarw site on January 30th, 1814 (Nance, *The Pottery and Porcelain of Swansea and Nantgarw* 1942), and presumably William Weston Young there agreed to pledge his commitment of £600 to the project, along with the £200 contribution from Billingsley and Walker granted to them by the Royal Worcester China Works prior to their departure. The lease for the Nantgarw residence and China Works site would have been arranged in late 1813 between William Billingsley and Edward Edmunds the landlord of the property. What is still unclear is when exactly Young met with Billingsley and Walker to set up the lease of the site, if, in fact the statement that he facilitated the introduction of Billingsley and Walker to Edmunds is correct? An earlier account suggests that Billingsley and Walker may have visited Lewis Dillwyn at the Swansea Pottery sometime late in 1813 to sound out the manufacture of porcelain there but Dillwyn was "not interested" in the approach. This is a curious statement chronologically because, although it would perhaps explain a possible involvement of Young in the negotiations, Dillwyn had already attempted to make porcelain at Swansea the year before, unsuccessfully, with the employment of two men for that particular purpose from the Coalport China Works (verified in a statement from Henry Morris, who worked at the Swansea China Works, in an interview with William Turner and recorded in turner's book, *The Ceramics of Swansea and Nantgarw* 1897) So, if that wish was still there, would it be reasonable to assume that Dillwyn have turned down the chance to employ Billingsley and Walker to remedy the situation just some months later? It transpires that even a short while later, after their failure to secure funding for the Nantgarw China Works through the *Memorial* application, Dillwyn was alerted to the opportunity afforded to recruit Billingsley and Walker to the fledgling Swansea China Works and he jumped at the chance! It seems therefore that we can dismiss the idea that he was not interested in recruiting them to Swansea in 1813 as rather fanciful.

References

H.G.M. Edwards, *Nantgarw Porcelain: The Pursuit of Perfection*, Series Editor: M.D. Denyer (Penrose Antiques Ltd. Short Guides, Thornton, 2017a). ISBN 978-0-244-90654-2

H.G.M. Edwards, *Swansea Porcelain: The Duck-Egg Translucent Vision of Lewis Dillwyn* (Penrose Antiques Ltd. Short Guides, Thornton, 2017b). ISBN 9780244325787

H.G.M. Edwards, *Swansea and Nantgarw Porcelain: A Scientific Reappraisal* (Springer, Dordrecht, 2017c)

H.G.M. Edwards, *Derby Porcelain, The Golden Years: 1780–1830* (Penrose Antiques Ltd. Short Guides, Thornton, 2017d). ISBN 9780244971663

H.G.M. Edwards, *Nantgarw and Swansea Porcelains: An Analytical Perspective* (Springer, Dordrecht, 2018)

H.G.M. Edwards, M.C.T. Denyer, *William Billingsley The Enigmatic Porcelain Artist, Decorator and Manufacturer* (Penrose Antiques Ltd. Short Guides, Neopubli, Berlin, 2016). ISBN 978-3-7418-6802-3

N.D. Gent, *The Patterns and Shapes of the Pinxton China Factory*, Non-Fiction Ceramics (1996)

J. Haslem, *The Old Derby China Factory* (George Bell, London, 1876)

W.D. John, *Nantgarw Porcelain* (Ceramic Book Co., Newport, 1948)

W.D. John, *Swansea Porcelain* (Ceramic Book Co., Newport, 1958)

W.D. John, *William Billingsley* (Ceramic Book Co., Newport, 1968)

A.E. Jones, Sir L. Joseph, *Swansea Porcelain: Shapes and Decoration* (D. Brown and Sons, Ltd., Cowbridge, 1988)

E.M. Nance, *The Pottery and Porcelain of Swansea and Nantgarw* (Batsford, London, 1942)

J. Robinson, R. Thomas, *Not Just a Bed of Roses: The Life and Work of the Artist, Ceramicist and Manufacturer, William Billingsley (1758–1828)* (Usher Gallery, Lincoln Publications, G.W. Belton Ltd., Gainsborough, 1996)

W. Turner, *The Ceramics of Swansea and Nantgarw* (Bemrose & Sons, Old Bailey, London, 1897)

J. Twitchett, *Derby Porcelain 1748–1848: An Illustrated Guide* (Antique Collectors Club, Woodbridge, 2002)

Chapter 8
The *Dinas* Refractory Silica Brick

Abstract From an historical introduction to brickmaking, the development of the Dinas refractory silica brick, its creation by William Weston Young and its usage in blast furnaces and ore smelting in the industrial revolution is reviewed. The singular properties of silica bricks and the interpretation of their properties from chemical analyses are discussed.

Keywords Dinas · Silica refractory brick · William Weston Young · High silica sandstone · Chemical analyses

8.1 History of Brickmaking

The history of brickmaking is almost ten thousand years old: the first bricks, made of clay or mud, sun-dried in simple moulds, dates from about 7500 BCE as found in Tell Aswad in Mesopotamia. Over the next thousand years of recorded history, ancient bricks of the same type have been found widely dispersed at excavations in Jericho, Catal Huyuk, Mohenjodaro and Harappa. The next stage involved ceramic bricks which had been fired at elevated temperatures in simple kilns and these have been located at Chengtoshan, China, dating from 4400 BCE: these bricks were fired in wood-burning kilns to 600 °C using red clay and were used principally for flooring materials. In the Quijaling period from about 3300 BCE, fired bricks were used in China to pave roads and in the construction of buildings. The Roman legions used portable and mobile kilns to make fired bricks in their military conquests, still using wood as a fuel, and they impressed the clay when wet with their legionary devices and motifs. It is believed that quite early on, the addition of some salt crystals to the kiln during firing gave a hardened surface to the bricks, but the salt glazing of fired bricks is generally more firmly attributed to the Chinese Song Dynasty around 1100 CE. Up until the mid-19th Century, bricks were hand-made in a range of dimensions until Henry Clayton patented the mechanisation of the brickmaking process in 1855, enabling some 25,000 bricks of a consistent size to be made in one day (Hudson, *Building Materials: Bricks and Tiles* 1972, pp. 28–42). At this point, the reader should be made aware of the term "firebrick", which is generically applied to bricks that have

© Springer Nature Switzerland AG 2019
H. G. M. Edwards, *Porcelain to Silica Bricks*,
https://doi.org/10.1007/978-3-030-10573-0_8

been made from fireclay, then fired in kilns and destined for high temperature work: confusion can be generated when authors use the special descriptor "silica firebrick", which actually, as will be seen in the discussion below, is a compositional nonsense in that silica refractory bricks contain little or no fireclay, unlike generic "firebricks", which can contain 30% or more clay. Another possible source of confusion is the name "firestone", which describes a type of natural stone much favoured for furnaces for melting brass and for the linings of lime kilns, in which a sandy loam is used as mortar, which forms a vitreous flux and covers the stone with a clear glass at high temperatures From the writings of Agricola (*De Re Metallica* 1556), Biringuccio (*Li Diece delle Pirotechnia* 1550) and White (*The Natural History of Selborne* 1813) it seems that the heat-resisting properties of this "firestone" were already recognised as early as the 14th Century.

Thus far we have considered the history of bricks which were essentially of one type, namely fireclay bricks: Jenkins (1942) has pointed out that the first special brick in history was the "*Windsor brick*", which was described in 1696 by a Swedish metallurgist, Thomas Cletscher, who visited the copper smelting works at Bristol and noted that the furnaces ....

> were built of ordinary bricks on the outside but on the inside they are lined with another kind of brick made from fire-proof earth and comes from Windsor. This kind is as good as it possibly can be and after a trial has been found to be better than the French. Similar earth has been found in Cornwall and in other parts of England.

This Windsor brick has been described as a soft red building brick made from alluvial sedimentary clays found in the Thames Valley, whose deposits have long been worked out and exhausted. Comprising about 70 parts sand and 30 parts clay, an analysis of surviving Windsor bricks carried out in 1848 (Knapp, *Chemical Technology* 1848, Vol. II, p. 481) yielded the following results: SiO_2 84.7%, Fe_2O_3 4.3%, Al_2O_3 8.9%, CaO 1.9% and MgO 0.4%. In 1725 another visitor to the Bristol smelting factory noted that they were then using Stourbridge bricks made from clay deposits found in the Midlands; it was thought that this clay may have been superior to that sourced at Windsor and was recommended for use in the prestigious glass making industries located in Stourbridge—a contemporary comment made in 1702 reads "The *glass making furnace being made of Stourbridge or Windsor brick ...*" (Savery 1702). Around the same time, Stourbridge clay was being made into bricks locally at the lead-smelting works in Neath by the Company of Mine Adventurers: two entries in their surviving log books read as follows: "*May 25th, 1703, Ordered that Mr Williams send some Stourbridge bricks ready-made, and some Neath Abbey coal to Mr Waller in Cardiganshire, by the same vessel; July 11th, 1704, Stores wanted at Neath—100 tons of Stourbridge clay*" (Grant-Francis, *The Smelting of Copper in the Swansea District from the time of Elizabeth to the Present Day: 1881* 1881, pp. 90, 92). As the furnaces were worked harder and used at higher temperatures their structures became more fragile and the need for a new type of refractory brick was evident.

The colour of bricks depends upon the source of the clay used in their composition and particularly on its iron (III) oxide content: generally, the chemical composition for "normal" fired bricks lies between the following rather narrow limits—sand (silica) 50–60%, clay (aluminosilicate) 20–30%, lime 2–5%, iron oxide <7%, magnesia <1%, and variations outside these limits, especially in the sand, clay and lime content, can produce seriously deleterious effects in usage on the fired brick, especially when a greater demand was placed upon it for increased furnace temperatures and resistance to vibrational shock.

8.2 Refractory Fire Brick

Unlike a normal brick, the refractory firebrick, or simply silica brick, has to withstand additional severe stress due to the elevated temperatures to which it is subjected in usage, in particular the thermal effects of expansion and contraction stresses induced by the heating and cooling cycles and their exposure to reactive flue gases and molten salts in the smelting kilns or furnaces. Thermal expansion in restricted furnace designs can place additional stress upon the brick lining structure and can result in crumbling or fracture of the linings, destroying the integrity of the bricks and thereby necessitating their replacement. It was found that the stability of the furnace bricks was critical for the proper operation of high temperature porcelain kilns as well as for the smelting furnaces—hard paste and soft paste porcelains require the maintenance of high temperatures in the range up to 1380-1420 °C for periods of up to one week at a time and metal-ore smelting furnaces can require even higher temperatures of perhaps up to almost 1700 °C.

The concept of the silica brick as envisaged by William Weston Young comprised a heat-proof brick using a mixture of very pure Dinas ground silica sand with the addition of 1% lime for binding the blue-grey Dinas rock powder, covered by his Patent No. 5047 in 1822. This meant that the whole brick body was silicified, unlike the patent (Patent No. 4168, of the 3rd October, 1817) previously taken out by William Harry of Swansea in 1817 for an early "*fire brick*" which used a composite of silica sand and water laid over normal brick furnace lining which at high temperatures melted to produce a thin veneer of molten silica on a normal ceramic brick. This apparently failed to fit the purpose adequately and in practice was less robust for use in the high temperature furnaces required for smelting iron ore and copper ores. Young realised that the absence of a binder was probably responsible and he trialled his silica brick experiments during 1820–1821 in the high temperature Nantgarw biscuit kiln erected earlier at the Nantgarw China Works site by Billingsley and Walker, when he finalised his recipe formulation for the large-scale production of *Dinas* refractory silica bricks in the Vale of Neath. It is interesting that the Nantgarw kiln was available to him for this experimental purpose – Billingsley and Walker having by then departed for Coalport in 1820—and the high temperature of firing required for the silica bricks was adequately encompassed by the Nantgarw kiln construction, which had consistently achieved estimated temperatures of 1400 ± 20 °C in the firing of the porcelain. This is a vindication of the kiln design and construction effected by Samuel Walker at Nantgarw, firstly in 1814. Owen (2014) has commented that Young's adoption of the term *"Dinas Firebrick"* to describe his invention and subsequent manufacture of the high silica refractory bricks from Dinas sandstone has been translated into several languages in industrialised nations, including Russian and Chinese!

An account of Young's successful invention of the refractory silica brick, which was apparently written by young himself in the mid-1830s, is reproduced below from Percy's book on metallurgy (Percy, *Metallurgy: The Art of Extracting Metals from their Ores* 1861):

> The material at the Dinas (the well-known rock of that name in the Vale of Neath), from which we procure it, is nearly pure silex; but, from its lying on the limestone and occasionally intermixing with it, there is taking the average of the general working, perhaps about 5 per cent of calcareous matter and 1 per cent of metallic, either iron or copper. Its use as a sand was discovered about 40 years ago (Author: this significantly predates William Harry's patent of 1817), when the fine of it was taken to one of the copper-works and used as a cement, and for mending their furnaces while at work, by placing it with a long iron-handled ladle or spade where the wash of the metal had destroyed the brick; and from its remarkable property of swelling in high heats, it fixed itself firmly. It gradually gained from one copper-work to another till its use became general: in fact, they are not able to find any other sand that will answer the purpose so well. Its fire-proof qualities being known, many attempts were made to produce a brick from it: but all the common combinations of different clays &c, failed. About 14 years ago (Author: i.e. about 1820) I became acquainted with it, and soon after devised a method of producing a brick from it of very extraordinary fire-proof qualities. When set in its own cement, for very high and long-continued heats it certainly will exceed in duration any other known brick. It does not suite every situation, as in fact, no fire-brick will; the nature of it at once tells you it must not be placed near alkaline substances; neither will the effluvia from some lead ores suit it. Perhaps it does not exceed Stourbridge [brick] for grates; but for the body of furnaces of most kinds it exceeds, as said before, that and every other known the brick in duration. The manner in which the brick is made gives it a rough coat compared with most others; indeed, it is peculiar in this respect; but as it is made in machines perfectly square, all eth works here prefer it with its rough coat; they say it sits better in the work, and they have now had more than 12 years' experience. This brick ought to be kept dry if possible for being open in its texture it imbibes moisture freely… The fire-place, roofs, sides and bridge of the furnace, also the lower part of the stack, should be built of Dinas; the back part and the remainder of the stack will do best of the other kind (similar to Stourbridge in quality); slabs for lining the flues and doors are also best made of this material. The appearance of the Dinas brick is peculiar in colour and for the roughness of its surface.

This was quite a discourse from Young to Percy, extolling the virtues, application and rather unusual surface roughness of *Dinas* bricks, but the reader will have noted that Young has carefully refrained from mentioning anything at all about his recipe or preparation process for the Dinas brick, which he worked in secret—as had Billingsley and Walker perfected in the Nantgarw China Works! In 1859, some 12 years after the death of William Weston Young, John Percy visited the Dinas Silica Brick Works, which was then under the supervisory management of Edward Young, nephew of William Weston Young, and his observations are illuminating and very precise: these record the earliest written record of the process of Young's invention and moreover provide the first description of the analytical chemical determinations of the Dinas rock starting material, but interestingly, not of the fired Dinas bricks product. Percy (*Metallurgy: The Art of Extracting Metals from their Ores* 1861) later recounted this information from his visit some two years later in his book:

> The mode of making the Dinas brick was long kept rigidly secret, and even now it is not generally known. The material, which is called "clay", is found at several places in the Vale of Neath, some of which I visited with Mr. E. Young (1859). It occurs in the state of rock and disintegrated like sand. Its colour, when dry, is pale grey. The rock, when not too hard, is crushed to coarse powder between iron rolls. By exposure to the air the hard rock becomes somewhat softer, but some of it is so hard that it cannot be employed. The composition of Dinas "clay" from two localities in the Vale of Neath is stated in the following table. The analyses were made in my laboratory by Mr. W. Weston. No. 1 was rock of medium hardness, which I obtained near Pont Neath Vaughan, on the occasion of my visit with Mr. E. Young; and No. 2 was sent to me from the same locality, though not from the same mine. (Author: these analytical determinations are presented in Table 8.1)

> The powder of the rock is mixed with about 1 per cent of lime and sufficient water to make it cohere slightly by pressure. This mixture is pressed into iron moulds, of which two are fixed under one press, side by side. The mould, which is open at the top and bottom, like ordinary brick moulds, is closed below by a moveable iron plate, and above by another plate of iron, which fits in like a piston and is connected with a lever. The machine being adjusted, the coarse mixture is put in the moulds by a workman, whose hands are protected by stout gloves, as the sharp edges of the fragments would otherwise wound them: the piston is then pressed down, after which the moveable bed of iron on which the brick is formed is lowered and taken away with the brick upon it, as it is not sufficiently solid to admit of being carried in the usual manner. The bricks are dried on these plates upon floors warmed by flues passing underneath; and when dry they are piled in a circular closed kiln covered with a dome, similar to kilns in which common fire-bricks are burned. About 7 days of hard firing are required for these bricks and about the same time for the cooling of the kiln. One kiln contains 32,000 bricks, and consumes 40 tons of coal, half free-burning and half binding. The price (1859) is 60 s. the thousand. They are manufactured of various shapes and sizes, to suit the furnace -builder.

> The fractured surface of one of these bricks presents the appearance of coarse irregular white particles of quartz, surrounded by a small quantity of light-brownish yellow matter. The lime which is added exerts a fluxing action on the surface of the fragments of quartz, and so causes them to agglutinate. These bricks expand by heat, whereas bricks made of fire-clay contract. On this account they are stated to be advantageous for the roofs of reverberatory furnaces, and in all parts where a solid and compact lining is needed. From their siliceous nature it is obvious that they should not be exposed to the action of slags rich in metallic oxides.

Table 8.1 Analyses of Dinas rock specimens (from Percy, *Metallurgy* 1861)

Analyte	Specimen No. 1/%	Specimen No. 2/%
SiO_2	98.31	96.73
Al_2O_3	0.72	1.39
Fe_2O_3	0.18	0.48
CaO	0.22	0.19
$K_2O + Na_2O$	0.14	0.20
H_2O	0.35	0.50
Total	99.92	99.49

It is intriguing that with such a detailed description of the processing of the Dinas rock and supporting batch on analytical data that almost 60 years later a scientific report of the geological strata of the Vale of Neath contained several errors intermingled with the facts as established by Percy. The *Memoirs of the Geological Survey for Merthyr Tydfil* (No. 231, 1904) quoted the following passage:

> The more highly siliceous bands in the Millstone Grit are quarried at Dowlais, in Penderyn-foel, at Craig-y-Dinas in the Neath Valley … for making the well-known Dinas fire-bricks. Sand derived from the disintegration of the mass of grit is dug for the same purpose at Pwll Byfre. At Dinas the best stone for making silica brick is known as "good-blue" and contains about 96–97 per cent of silica, with not more than ½ percent of lime or ½ per cent of iron, the remainder being alumina. It has a pale-blue tinge and breaks down readily into powder. After being crushed it with milk of lime, moulded and burnt for six days, at a temperature which is said to reach 4000 °F. The stone is a conglomerate of vein-quartz in a grit-matrix, but the pebbles break down as readily as the matrix.

At first sight this reads well and seems to echo the notes that Percy made after his visit to the Dinas Silica Brick Company in 1859, except for the quoted firing temperature for the *Dinas* bricks stated so authoritatively as 4000 °F, which equates with 2205 °C—this is a phenomenally large temperature compared with those used at the Nantgarw China Works by Young in his trial experiments, which approached 1420 °C! It is inconceivable that any furnace or kiln should operate at a temperature in excess of 2000 °C in 1820—let alone maintain this at a steady temperature for six days. We must therefore treat this quoted temperature as a fictional entity, although the remainder of the article seems to be quite correct in its evaluation of the Dinas rock composition as we can see from Table 8.1.

In just a few years after Young's factory had commenced the manufacture of silica bricks, several others were set up locally and elsewhere to satisfy the growing demand for this material in an industrialised society: an example is illustrated in Fig. 8.1, showing an advertisement for the Penwyllt (literal translation: *wild headland*) Silica Brick Works at Craig-y-Nos, just a few miles away from Dinas Rock, which continued operating until the 1930s. Note that the Penwyllt advertisement labels its product as *Dinas* silica bricks to avoid confusion with the less effective clay firebricks. The news of the effectiveness of the silica brick for furnace construction in a heavily industrialised society soon spread widely.

Fig. 8.1 Advertisement of the Penwyllt Silica Brick Works Co Ltd. products, Neath, South Wales. Note that they advertise their product as Dinas Silica Brick to avoid confusion with "firebrick"

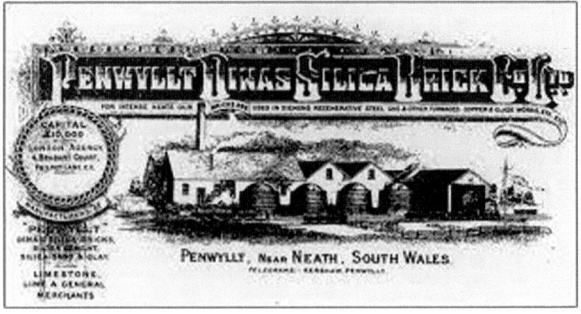

In the Appendix of the book inserted by the translators of Knapp's *Chemical Technology* of 1848, under the heading "*Stone-bricks*" (which is itself worthy of note as a new variant on the nomenclature of *firebrick* and *firestone*, mentioned above) the following statement is made:

> There is a very peculiar brick, manufactured at Neath in Glamorganshire... which has been found very valuable in constructing furnaces, more particularly the arches of the reverberatory furnaces employed at Swansea in melting copper ores. There cannot be a doubt that, for the construction of many descriptions of chemical furnaces, such bricks would prove of essential service.

Jenkins (1942) amplifies this statement from a century before by highlighting the continued use of the silica brick for steel-making in most parts of the world and he comments that although other additives have been tried, the lime binder adopted by William Weston Young in 1821 for his *Dinas* silica brick is still the favoured material in general use for the manufacture of silica bricks almost two hundred years later!

As mentioned above, several contemporary statements refer to the continuation of porcelain manufacture at Nantgarw after the departure of Billingsley and Walker for Coalport in 1820 and it could be reasoned that this referred to Young's firing up the kiln for his brick formulation recipes and that, in fact, no porcelain was made thereafter in Nantgarw and the actual operation there was exclusively that of decorating already glazed porcelain and the glazing and decoration of biscuit porcelain remaining in stock in the white by Young and Pardoe. The firing of the kiln by Young for his experiments on silica brick production could well have been misinterpreted here for his attempts at porcelain manufacture, but we also know from his Diary entries that Young already had the serious intention of reviving porcelain production at Nantgarw and was also working on a porcelain formulation at the same time as he was inventing the silica firebrick. A report in Nance (*The Pottery and Porcelain of Swansea and Nantgarw* 1942) mentions that in one of the several interviews with ex-employees of the Nantgarw China Works conducted by William Turner for his earlier seminal work (*Ceramics of Swansea and Nantgarw* 1897), a child worker during Young's tenancy between 1820 and 1823, Richard Millward, commented that he did make porcelain there and that the kiln wastage was very high. Morton Nance has concluded that this indicated that Young must have had access to the Billingsley/Walker Nantgarw porcelain formulation either directly on their departure or from Lewis Dillwyn, but we now realise that this cannot have been correct as Young's entries relevant to his porcelain experiments in his Diaries for 1819/20 (Young, *Fact-Books*) do not include a bone-ash component, which was so essential for the translucency of the Nantgarw fired body and also for the Swansea duck-egg porcelain translucency. We can therefore assert quite definitively that Young cannot have received any information about Billingsley and Walker's Nantgarw formulation and recipe for his own experimental recipes—which as we shall show are indicative of Young manufacturing a hard paste porcelain version.

8.3   Dinas Rock Sandstone

Craig-y-Ddinas (Welsh: *fortress rock*) is actually a carboniferous limestone promontory (Fig. 8.2) situated near the confluence and between the Afon Mellte and Afon Sychryd tributaries of the Afon Nedd in the upper Vale of Neath in South Wales, an area of spectacular natural beauty, as written by Lewis Dillwyn, William Weston Young and Alfred Russel Wallace, with impressive waterfalls now part of the designated Fforest Fawr UNESCO Global Geopark. It derives its name from the Iron Age earthworks discovered at its summit—Dinas is a Welsh word for a defensive site. A very high purity, almost 100% SiO_2, silica rock was discovered there in 1780 and was mined underground through several adits driven through the side of the hill; these still exist, although they are now in a highly dangerous condition and access is not advisable without taking adequate precautions. Tramways for horse-drawn trucks and aerial cableways were constructed through this difficult terrain to transport the silica rock to the adjacent Neath Canal as the River Neath (Afon Nedd) is not navigable at this point. It is recorded that an earlier ceramic or brick factory was established at a site near Pont Walby by Fredericks and Jenner, which could have possibly been taken over by Young and his associates in 1821 to eventually make their *Dinas* silica bricks, although this is merely conjectural. The established Dinas Silica Brick Works was initially very successful but closed eventually in 1904 and operations were then transferred to Swansea. At the site, industrial archaeology has revealed the remains of four circular limekilns and sand pits, with ruins of processing buildings—many of the mine galleries are now in a dangerous condition, are fully

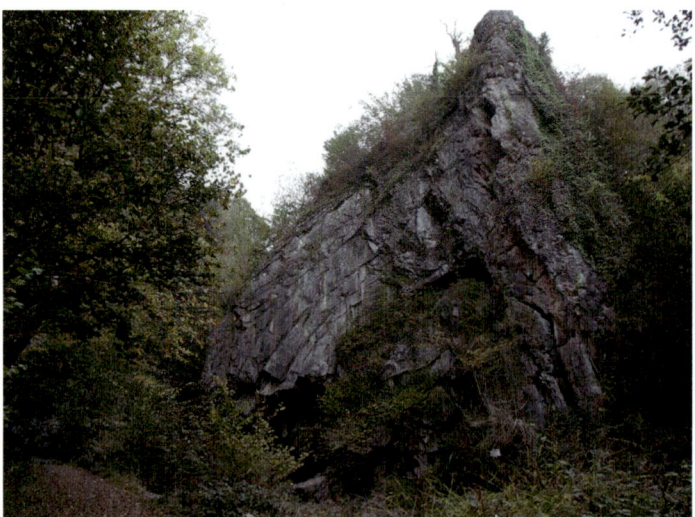

Fig. 8.2 The sandstone outcrop known locally as Bwa Maen, near Pont Walby, Vale of Neath, landmark for the Dinas rock silica mine. This has frequently been incorrectly described as the "*Dinas Rock*"

or partially flooded, and rusty remains of abandoned trucks, rails and tramways can still be seen. It has been recorded that the extraction of the silica rock was undertaken manually and slowly, and that care needed to be taken in its transportation and preparation to minimise the effects of the fine silica dust on the workforce (resulting in lung disease such as "silicosis", caused by the pronounced breathing of aerially transported very fine silica particles). Hence, the use of gunpowder for blasting was not encouraged.

8.4   The Refractory Silica "Firebrick"

William Weston Young is credited with the invention and the first manufacture of a "silica firebrick", or more properly, a silica brick, designed to withstand the very high temperatures operating in brick-lined kilns and furnaces for iron and copper ore smelting, coke ovens and in glass production. To achieve this, he utilised a particularly pure form of silica excavated at Dinas Rock (Craig-y-Ddinas) in the upper Neath Valley, which saw the first of his silica bricks being produced commercially in 1822 after the success of his initial trial experiments which he carried out on the Nantgarw China Works site. The key raw material is the very high purity silica rock mined on site with very few impurities present.

The silica rock found at Craig-y-Ddinas was an exceptionally pure form and analytical figures quoted nearly a century later (see Bywater, in Ross, *National Bureau of Standards No. 116* 1929) give the following composition for Dinas rock: silica 97.6%, alumina 0.5%, iron oxide 1.5%, calcium oxide 0.2%, potassium oxide 0.1% and soda 0.03–65% of the silica grains were of a diameter 0.2 mm, 25% between 0.2 and 0.4 mm, 5% less than 0.2 mm and 5% more than 0.7 mm in size. The Dinas rock silica is a pale greyish-blue colour: after manual excavation it was ground and mixed with 1–1.5% lime and a *very small amount* of silicaceous clay to harden the mixture, water was then added to fluidise the mixture for its formation into bricks, these were then dried at 45 °C and then finally fired at 1450 °C or higher to produce the finished silica brick. For a time, there was no alternative source of such a fine silica raw material, until a black *ganister* was later found in the lower coal measures at Sheffield with a similar composition to that of Dinas rock. It is interesting that this ganister was used as a local road bed material in Sheffield and after its disintegration under heavy traffic loads the powdered siliceous material was collected and used as effective refractory cement in the steel-making furnace linings of local industries as a substitute for the *Dinas* fire bricks (Jenkins 1942).

The formation of a refractory silica brick at a high temperature is chemically dependent upon several factors, the most important of which is the conversion of the low-temperature stable alpha-quartz into the high-temperature polymorph, beta-quartz, which occurs thermally at 575 °C, and then its further polymorphic high temperature conversion into, firstly, beta-tridymite at 870 °C and then finally into beta-cristobalite at 1470 °C. Side reactions of the additives or impurities present are very important: for example, the lime, calcium oxide, reacts with silica at 1200 °C to form beta-wollastonite, $CaSiO_3$, and in the presence of iron oxide impurity, calcium ferrite, $CaFe_2O_4$, and hedenbergite, $CaFe(SiO_3)_2$, are formed (Edwards, *Nantgarw and Swansea Porcelains: An Analytical Perspective* 2018). In the high temperature mixture, the silica matrix, which analyses chemically as SiO_2, comprises four main components: quartz, cristobalite, tridymite and a silicaceous glass cementing these together, assisted by the alkaline calcareous flux—the matrix is grainy, and coarse grains are composed primarily of cristobalite, whereas the fine grains are comprised of tridymite, wollastonite and the silica glass. In a good, well-formed silica brick, therefore, little quartz remains unchanged during the firing process, although chemical analyses still show the presence of silica, SiO_2, since the cristobalite and tridymite polymorphs are analytically of identical chemical formulation. A typical mixture for a good silica brick "paste" is given generically as silica sand (97%), lime (2%) and water (1%).

Several review articles specify the requirements for the composition and comparative evaluation of silica bricks (Alnawafleh 2009; Mohanty et al. 1986; Chester, *Refractories, Properties and Their Applications* 1973; Ross, *National Bureau of Standards* 1929; Insley and Klein 1919) and a summary can be provided as follows. The basic requirement is for the silica brick to withstand high temperatures up to 1600–1700 °C after firing: the melting points of the individual high temperature components of the matrix are quartz at 1725 °C, cristobalite at 1710 °C and tridymite at 1670 °C. Because of the differential densities of quartz (2.65 g cm^{-2}), cristobalite (2.31 g cm^{-2}), tridymite (2.27 g cm^{-2}) and the fused calcareous silica glass matrix (2.20 g cm^{-2}), the density changes upon initial firing of the brick produce a significant change in volume of the specimen amounting to about 14–15% and as a result the temperature ramp (rate of increase of temperature with time) of the kiln upon firing needs to take place slowly, with a heating rate not exceeding 5 °C min^{-1}. The efficiency of conversion of the quartz raw material into its high-temperature polymorphs cristobalite and tridymite is dependent upon this rate of heating, the final temperature attained and the time spent in the kiln at the highest temperature: the bricks are maintained at a final temperature of 1470–1500 °C for about 36–48 h before cooling is effected at an equally slow rate to permit a volume decrease of about 4% to be accommodated without incurring internal fracture zones and stress cracks within the fired brick which would be sites for its potential future structural weakness. The final density of a well-fired silica brick would approach the value of 2.33 g cm^{-2}, within which the residual quartz component is only estimated at 6%, the composition of the brick then being substantially that of cristobalite and tridymite, both chemically represented as SiO_2. A typical chemical composition of a well-fired silica brick at room temperature would thus be 6% alpha quartz, 36% cristobalite, 45% tridymite and 13% silica glass matrix. Small quantities of impurities will suffice to colour the brick, the most common being iron (III) oxide. A significant increase in the proportion of lime in the initial mixture does seem to have an effect upon the refractory properties of the brick as the diffusion of calcium ions into the quartz structure at a high temperature of around 1250 °C assists the formation of tridymite (Mohanty et al. 1986). However, the alumina content, which is found through the incorporation of clay into the initial mixture (clay is a sodium aluminosilicate, of which some 25%

is composed chemically of Al_2O_3), drastically affects the integrity of the silica brick and especially the liquid phase at high temperature which can result in a load creep under stress, which is not a desirable proposition for kiln construction. For example, if the alumina content of the raw materials is 1% or more, the high temperature limit of the silica brick usage is lowered by about 50–100 °C or more, compared with an alumina content in the standard brick of only 0.5% or less.

The integrity of a well-fired refractory silica brick, which should have a compression strength of approximately 300 kg cm^{-2} in the direction of the compressive load, is preserved during normal use up to its softening point of 1700 °C; the greatest source of structural failure of a silica brick during use is thermal shock to which it has been exposed in multiple heating/cooling cycles in the kiln or oven: tests carried out on sample bricks hence involve the repeated heating of specimens to 900 °C followed by their sudden cooling in ice-water—a good silica brick should be able to withstand more than ten such drastic thermal cycles without exhibiting visible deterioration.

In this context, it is rather incongruous to find that some earlier historical authors have asserted that William Weston Young used a significant proportion of clay in his *Dinas* silica brick construction: the evidence for this assertion is obscure but a figure of up to 30% has been cited for the so-called recipe, which would have corresponded to an actual alumina content of some 7–8%. The reason for this could possibly be attributed to the confusion arising from the use of the term silica firebrick for the silica brick, as firebricks generally contain significant quantities of sodium aluminium silicate, or fireclay—it would perhaps to be reasonable to infer that a silica firebrick, *aka* silica brick, would therefore contain a significant proportion of fireclay, which it does not! Owen in 2014, citing the *British Clayworker* manual of 1907/8, has commented that William Weston Young's raw material composition in his formulation for silica bricks was: 95% silica, 4% clay and 15 fluxing oxides, which would in itself be a most unlikely adulteration of the desirable Dinas rock component undertaken by Young and would have given an undesirably large alumina content, detracting from the integrity of the Dinas silica brick formed from this recipe.

We can see from the above discussion that these high alumina figures would inevitably have resulted in a highly imperfect refractory brick, particularly with respect to its high temperature capability of usage in smelting furnaces for which it would have been specifically designed by Young. This is an apparent paradox as it is clear that the incorporation of such a recipe would certainly have not enabled Young to have maintained his global market lead in the manufacture of the highest quality refractory bricks for several decades—it seems superfluous to add that Young's delight in securing the highly pure Dinas silica rock (~98% silica) would have been negated by his then mixing that with up to 30% clay as had been suggested, in a 70:30 mixture, although his stated 1% lime additive to this mixture would have been actually quite correct according to modern analytical experimentation! In fact, the 70% sand and 30% clay that has been proposed for Young's refractory silica brick seems very close to the composition of an ordinary house brick (a firebrick) and is also very close indeed to that of Chinese hard paste porcelain!! A possible explanation could be that Young's recipe for his hard paste porcelain furnished in his diary notes in 1819 was confused by some later chroniclers with his trial experiments on refractory bricks, both of which were seemingly fired at the Nantgarw site in the biscuit kiln, especially when others have maintained categorically that no porcelain was made at Nantgarw after the departure of Billingsley and Walker? Another possible mis-interpretation would occur by Young's use of the word "fireclay" in some documentation for his composition of the Dinas firebrick in place of the silica rock—and fireclay is a geological term that was applied to aluminosilicate clay formations for brick construction. Of course, the Dinas silica brick does not really contain clay at all and the very high purity silica of which the formulation is composed specifically excludes this component—also his 1% lime additive is just right for the silica binding needed at high temperatures. Another possibility, less likely perhaps but which would also generate confusion, is the adoption of the term calcium silicate brick, which sometimes also operates under the misnomer "silicate firebrick", for a silica firebrick or silica brick: the two are very different products and the silicate brick, which does contain a significant clay to silica ratio has often been chosen for its resistance to frost damage in low level building projects but it certainly does not merit an implied association with a true firebrick as it does not have its high temperature properties.

8.5   An Analytical Theme

It has been recorded, it is now believed incorrectly, that William Weston Young's *Dinas* heat-resistant refractory silica bricks could have comprised 70% silica, 30% clay and some 1% lime additive to increase the binding of the two components when fired at high temperatures. This conclusion made by some earlier authors, it is now believed incorrectly, probably arose from their belief that the manufacture of a silica brick involving almost pure silica as its primary component was not realistic as the high temperatures required for the melting of pure silica could not be achievable at that time without the significant addition of a binder and alkaline flux. When the composition of Chinese hard paste porcelain is compared, a rather remarkable coincidence is perceived: in 1922, Herbert Eccles and Bernard Rackham undertook the wet chemical analysis of porcelain specimens from the Victoria and Albert Museum, South Kensington, London, which included Chinese hard paste porcelain from the K'ang Hsi period (1662–1722) and a range of English and Welsh soft paste and hard paste porcelains, including Nantgarw and Swansea (Eccles and Rackham, *Analysed Specimens of English Porcelain* 1922). This was the first systematic chemical analysis of English and Welsh porcelains to have been undertaken and in our context the comparative results are revealing. Table 8.2 shows the major chemical components of the *Faux Dinas* brick and Chinese hard paste porcelain to illustrate how this possibility of confusion arose. Table 8.3 illustrates the chemical composition of several hard paste and soft paste porcelains (including those of Swansea and Nantgarw) and that of the real Dinas silica bricks for comparison:

Table 8.2 Comparative chemical analyses of porcelains (%) and *Faux* Dinas silica bricks

Specimen	Silica	Alumina	Lime	Phosphate	Potash and soda
Faux Dinas silica brick	70	29	1		
Chinese porcelain[a]	72	23	1		4
Bristol porcelain[a]	70	24	2		4
New Hall porcelain[a]	70	24	2		4
Nantgarw porcelain[b]	46	17	20	14	3
Swansea porcelain[b]	48	26	13	10	3
NG4	68	21	4	4	3
NG6	80	13	1		4
NG7	66	18	6	8	2

[a]Hard paste porcelains
[b]Soft paste porcelains

Table 8.3 Comparative analytical data for elemental oxide composition of hard paste Chinese (K'ang Hsi) and soft paste Nantgarw (Billingsley and Walker) porcelains

Elemental oxide	Hard paste Chinese[a] (%)	Soft paste Nantgarw[a] (%)
Silica	72	46
Alumina	23	17
Lime	<1	20
Phosphate	<1	14
Magnesia	<1	<1
Potash	2	3
Soda	2	<1
Lead oxide	<1	0

[a]From Eccles and Rackham (1922): K'ang Hsi porcelain (1682–1722), Nantgarw (1817–1819). Elemental oxide data rounded off to nearest %

It can be concluded that, according to earlier incorrect formulations, the *Faux Dinas* refractory brick composition would have been very close indeed to that of hard paste (true) porcelain whereas the soft paste porcelain, which contains a significant proportion of bone ash (calcined calcium hydroxyapatite) in its recipe, namely about 40% for Nantgarw and 25% for Swansea porcelain, does not have an analytical match at all. In the analytical figures shown here, silica represents the silicaceous component, represented by Dinas rock for the bricks and fine quartz river sand and flints for the porcelains, alumina represents the clay component (clays are sodium aluminosilicates), phosphate represents the calcined bone ash component, lime represents the sintered calcite or limestone component and the residual potash and soda represents the potash added and soda components of the clays, which together with the lime provide the alkaline flux assisting the component fusion at high temperatures. Of course, as it is now realised, the real *Dinas* refractory silica brick formulation differs significantly from that believed to be represented by the *Faux* brick, which better describes the hard paste porcelain perfected by Young at Nantgarw in 1819 and 1820.

A very interesting situation can now be postulated: it is known that William Weston Young did undertake experimentation in the composition of his silica bricks at Nantgarw in 1819 and 1820 which he then fired there in the high temperature porcelain kiln, which was capable of reaching a temperature of at least 1400 °C. Although he would not have had the analytical figures of Eccles and Rackham available, as these did not appear until almost exactly one hundred years later in 1922, he would nevertheless surely have been aware of the composition of Chinese porcelains as the recipe and formulation which had been a closely guarded secret for hundreds of years was revealed by the Jesuit priest, Father Francois Xavier d'Entrecolles, to his superiors in Paris in 1729 through hearing confessional from his Chinese converts to Catholicism who worked in the porcelain industry. This knowledge supported the significant and advantageous growth of the French porcelain industry in the mid-18th Century and the proliferation and growth of smaller French porcelain manufactories prior to the French Revolution in 1789 (Edwards, *Nantgarw and Swansea Porcelain: An Analytical Perspective* 2018). The ability of the Nantgarw kiln to fire the *Dinas* brick formulations could well have initiated a thought to undertake porcelain manufacture again at Nantgarw after Billingsley and Walker had departed for Coalport early in 1820 and left Young with a large stock of Nantgarw porcelain in the white for decoration and subsequent sale at the auctions in 1822 and 1823. The statement has often been made that porcelain production at Nantgarw ceased with the departure of Billingsley and Walker in 1820 and this would at first seem to be a reasonable presumption since the brains and drive behind the recipe and its firing and the essential expertise in kiln management was no longer perceived to be present there. However, Young would have been involved, even perhaps peripherally, in the stages of the production process and his successful experiments in devising the high-silica Dinas refractory brick lends testimony to his ability to create novel commercial ceramics. It is entirely logical to suggest therefore that Young may well have contemplated the synthesis of a hard paste porcelain in Nantgarw in 1820 and the years immediately following? The robustness and hardness of hard paste porcelain fired at similar temperatures to those required for his Dinas refractory silica bricks would not have been lost on Young and he would have been in rather desperate financial circumstances which could possibly have catalysed and initiated trials of this sort, having taken on the failed Nantgarw China Works operation in 1820. This is not such a fanciful hypothesis, as recent analytical studies of porcelain shards excavated from the Nantgarw waste pit on the Nantgarw China Works site have revealed the presence of a hard paste porcelain there. Its presence there is completely unexpected—as it is known that Billingsley and Walker during their careers never ventured into the manufacture of this type of porcelain and we need to explain its presence at the Nantgarw site!

The specimens NG4, NG6 and NG7 given in Table 8.2 are Nantgarw porcelain shards excavated from the China Works site and subjected to the most recent analytical study (Edwards and Colomban 2019) involving elemental and molecular spectroscopic determinations which will be discussed in detail later, along with other shards from the same site. Whereas most of the shards provided for chemical analysis conform to the established bone ash recipe for soft paste porcelain quoted by Taylor (*The Complete Practical Potter* 1847) which was apparently provided to him by Samuel Walker just before he departed for the USA, and all therefore contain a component of calcined bone ash representing about 40% composition from the phosphate analyses given, the three NG shards in Table 8.2 are clear exceptions to this recipe. In particular, shard NG6 is better described as a "hard paste porcelain" in classification and matches closely those cited in the same Table for Chinese and English hard paste porcelain factories, with no presence of phosphate, and hence calcined bone ash, at all. The other two shards, NG4 and NG7, are best described in Eccles and Rackham's classification as "hybrid porcelains", in that they contain some semblance of the properties of both extremes. This is not the first time that the analysis of Nantgarw porcelain shards has raised a query about the possibility that the Nantgarw China Works did create a novel porcelain that was not true to type, as Victor Owen in the 1990s found a similar occurrence in several siliceous shards from an archive at the Nantgarw China Works site (Owen and Morrison, 1999). This gave rise to a conclusion that perhaps the Nantgarw China Works did experiment with a different type of porcelain; however, several explanations should be considered for the presence of a hard paste porcelain type shard in the waste dump at the Nantgarw China Works site:

An possible explanation for the presence of shard NG6 is that Billingsley and Walker purchased "grog", a batch of porcelain wasters from another factory site of indeterminate origin: this practice has been recorded in other porcelain manufactories such as Derby, where William Duesbury in 1795 bought consignments of "grog" amounting to several tons weight from London, which he then ground up and fired in admixture with his own porcelain at the Derby China Works (Anderson, *Derby Porcelain and the Early English Fine Ceramic Industry* 2000). Clearly, if this "grog" contained hard paste porcelain then the analytical composition of the believed source factory would be anomalous, as seen in our Nantgarw examples. There is no record whatsoever that Billingsley or Walker did this at Nantgarw or indeed at any other factory where they had been employed hitherto: the purist, Billingsley, would certainly not have favoured this practice as he was in a mindset to produce objectively and without compromise the best and most translucent porcelain, which would not have been served by the introduction of "grog" of variable composition from another factory.

Because the firing of Nantgarw porcelain resulted in appallingly high losses in the kiln, estimated at around 90%, through sagging or warping attributed to the body formulation and also perhaps to ineffective temperature control, it could be that Billingsley and Walker in attempting to deliver a service commission to a tight deadline would have purchased items from another factory to complete the service. This happened elsewhere in other factories, and a noteworthy case is provided by the neighbouring factory at Swansea, where Lewis Weston Dillwyn bought in some Coalport pieces in the white for the completion of his prestigious *Lysaght* service (Fig. 8.3), all decorated at Swansea by Henry Morris and all marked SWANSEA in red enamelled stencil (Edwards, *Swansea and Nantgarw Porcelains: A Scientific*

Fig. 8.3 Swansea porcelain
plate from the prestigious
Lysaght dinner-dessert
service made in Dillwyn's
famed duck-egg translucent
porcelain, ca. 1817–1819.
Decorated locally by Henry
Morris with a central basket
of flowers, profuse gilding
and with a cobalt blue
ground colour at the verge. A
pair of these plates were
presented to Her Majesty
Queen Elizabeth II upon her
accession to the throne in
1953 by the Swansea
Corporation on behalf of the
people of Swansea. Private
collection

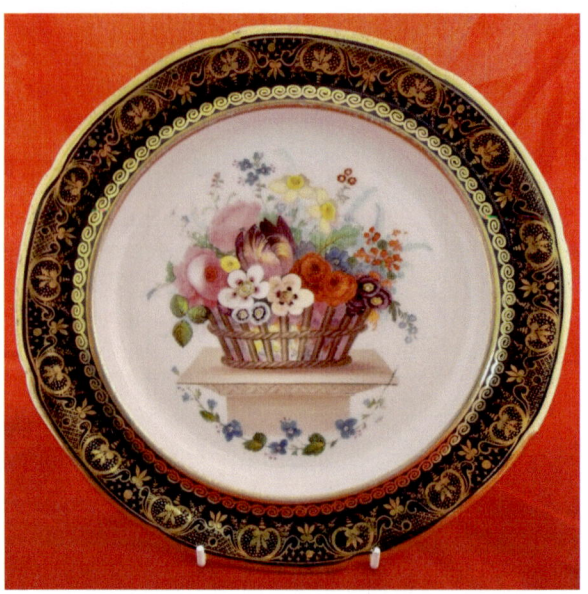

Reappraisal 2017). The famed Swansea soft paste porcelain of this period had a
characteristic duck-egg translucency perfected by Dillwyn and Walker in 1815 and
the Coalport soft paste is not too dissimilar—but some non-characteristic Swansea
shapes in the *Lysaght* service can certainly be attributed to the Coalport factory, such
as dessert comports, although the plates are all of the same size and shape. Another
possibility for such a mixed Swansea-Coalport service arises in the *Biddulph* service,
made for Lord Biddulph in 1817 and decorated by Philip Ballard in the London ateliers, where it has been suggested that several pieces originated in Coalport. An even
more curious combination is that of a mixed Swansea and Nantgarw tea service made
for the Marquess of Anglesey, where the characteristic London-shaped Swansea tea
cups are matched with those of Nantgarw of different shape—but the decoration on
all pieces is identical: a distinct possibility here is that there were in fact two tea
services ordered by the Marquess of Anglesey, one from Swansea and the other from
Nantgarw, both having identical decoration! The absence of factory order books at
Nantgarw and at Swansea does not assist in ascribing a correct forensic attribution
and reliable provenance for these items.

Another explanation for the discovery of hard paste porcelain shards in the waste
pit of a factory which hitherto has always been regarded as an exclusive manufacturer of soft paste porcelain is that some clandestine, unrecorded experiments did
take place to change the composition of the porcelain body. The release of the Nantgarw formulation to John Taylor in 1847 (and published in his book, *The Complete
Practical Potter* 1847), apparently from Samuel Walker, does not in itself mean that
there had not been attempted another perhaps inferior formulation at Nantgarw which
had then been discarded. Whilst at Swansea, Billingsley would have had knowledge

of Dillwyn's attempts to improve the robustness of his china body through a series of experiments which Dillwyn carried out and documented fully between 1815 and December 1817 (see Appendix B for a Transcript of Dillwyn's Notes and also Eccles and Rackham, *Analysis of English Porcelains* 1922). Billingsley had left Swansea to set up again the second phase at Nantgarw towards the end of these empirical trials, but it is interesting that Walker remained to assist Dillwyn in his endeavours until September 1817. Eventually Dillwyn did make a more robust porcelain body which he termed his *"trident"* paste, but this had lost the superb duck-egg translucency of his former body and had instead a muddier, more diffuse translucency with a rather inferior and unpleasant pigskin-like surface texture. Despite being decorated in the finest fashion by Swansea artists such as Henry Morris and David Evans, the trident paste was distinctly out of favour with a London society more used to the extremely fine and delicate duck-egg variety, and as a result Dillwyn lost his market edge rapidly and was faced with bankruptcy in less than two years. The question remains: did Young attempt to make a more robust version of Nantgarw porcelain at Nantgarw in 1819/1820: Walker departed Nantgarw for Coalport along with William Billingsley.... leaving perhaps the seeds of the idea with William Weston Young, already embarking upon his Dinas firebrick experiments at Nantgarw? It is clear that documentation exists in Young's Diaries which definitively indicates that he undertook some attempts to manufacture a different type of porcelain at Nantgarw after Billingsley had left in 1819 and 1820. It is certainly recorded that after the death of William Billingsley in Coalport in 1828, Samuel Walker approached Dillwyn at Swansea with the suggestion that he may consider the start-up again of porcelain manufacture at either Swansea or Nantgarw, but Dillwyn had had enough of porcelain production at this stage and he had already passed on the reins of his ceramic manufacture at Swansea to his son, Lewis Dillwyn Jr. It is interesting to speculate further on the comment that has been perpetuated, seemingly incorrectly as it now transpires, that in the auctions of 1822 and 1823, the hardware, china moulds and kilns had been purchased from both Swansea and Nantgarw and dispersed—the main recipient being named as John Rose of the Coalport China Works. Billingsley and Walker were at that time both in employment at Coalport and it has been stated by John Rouse, an employee at the Coalport China Works in the 1820s, that they were engaged there in manufacturing Nantgarw porcelain body compositions for Coalport, which were "sold as such". This may not be without substance. but in another context both are mentioned as giving artistic advice in the enamelling workshops and even, in Walker's case, the invention of a beautiful purple ground colour which the Coalport China Works became much endeared to.

References

G. Agricola, *De Re Metallica, Libri XII* (J. Froben & N. Episcopius, Basel, 1556)

M.A. Alnawafleh, Mechanical and physical properties of silica bricks produced from local materials. Aust. J. Basic Appl. Sci. **3**, 418–423 (2009)

J.A. Anderson, *Derby Porcelain and the Early English Fine Ceramic Industry*. Ph.D. Thesis, University of Leicester, UK, October, 2000

V. Biringuccio, *Li Diece Libri Della Pirotechnia, Nelli Quali Si Tratta non Solo la Diursite Delle Minere*, 1st edn., P. Gironimo Gigli, Venice, 1550 (MIT Press, Cambridge, 1996)

H. Chester, *Refractories, Properties and their Applications* (Iron and Steel Institute, London, 1973)

L.W. Dillwyn, *Notes on the Experimental Production of Swansea Porcelain Bodies and Glazes*. Made by Lewis Weston Dillwyn with Samuel Walker at the Swansea China Works Between 1815 and 1817. Presented to the Library of the Victoria &Albert Museum, South Kensington, London by John Campbell in 1920. Reproduced in Eccles & Rackham, *Analysed Specimens of English Porcelain*, 1922, see reference below

H. Eccles, B. Rackham, *Analysed Specimens of English Porcelain in the Victoria and Albert Museum* (Victoria and Albert Museum, London, 1922)

H.G.M. Edwards, *Swansea and Nantgarw Porcelains: A Scientific Reappraisal* (Springer, Dordrecht, 2017)

H.G.M. Edwards, *Nantgarw and Swansea Porcelains: An Analytical Perspective* (Springer, Dordrecht, 2018)

H.G.M. Edwards, P. Colomban, to be published (2019)

G. Grant-Francis, *The Smelting of Copper in the Swansea District from the Time of Elizabeth to the Present Day:1881* (H. Sotheran & Co., London & Manchester, 1881)

K, Hudson, *Building Materials: Bricks and Tiles* (Longmans, London, 1972), pp. 28–42

H. Insley, A.A. Klein, Constitution and Microstructure of Silica Brick and Changes Involved through Repeated Burnings at High Temperatures, *Department of Commerce Technologic Papers of the National Bureau of Standards,* No. 124 Government Printing Office, Washington, July 11, 1919

R. Jenkins, The silica brick and its inventor, William Weston Young, 1776–1847. Trans. Newcomen Soc. **XXII**(1), 139–147 (1942)

F.L. Knapp, *Chemical Technology: Chemistry in its Application to the Arts & Manufactures*, ed. and trans. by T. Richardson, H. Baillie (London, 1848)

P. Mohanty, N. Singh, D. Smith, Silica—a critical study. Refract. J. **12**, 5–9 (1986)

E.M. Nance, *The Pottery and Porcelain of Swansea and Nantgarw* (Batsford, London, 1942)

M.J. Owen, The Dinas silica brick. Silicon **6**, 91–92 (2014)

J.V. Owen and M.L. Morrison, Sagged phosphatic Nantgarw porcelain (ca. 1813–1820): Casualty of overfiring or a fertile paste. Geoarchaeology, **14**, 313–332 (1999)

J. Percy, *Metallurgy: The Art of Extracting Metals from their Ores, and Adapting Them to Various Purposes of Manufacture of Fuel, Fire Clays, Copper, Zinc, Brass, Etc.* (J. Murray, London, 1861)

D.W. Ross, Silica Refractories: Factors Affecting their Quality and Methods of Testing the Raw Materials and Finished Ware, *Department of Commerce Technologic Paper of the Bureau of Standards,* No. 116, Issued April 19th, 1929 (Government Printing Office, Washington, USA, 1929)

T. Savery, *The Miner's Friend: Or, An Engine to raise Water by Fire Described and of the Manner of Fixing it in Mines* (S. Crouch, London, 1702)

J. Taylor, *The Complete Practical Potter* (Shelton, Stoke-upon-Trent, 1847)

W. Turner, *The Ceramics of Swansea and Nantgarw* (Bemrose & Sons, Old Bailey, London, 1897)

G. White, *The Natural History and Antiquities of Selborne and a Garden Kalendar* (White Cochrane & Co., London, 1813)

W.W. Young, *The Diaries of William Weston Young, 1776–1847 (1802–1843)*, 30 vols., West Glamorgan Archive Service, Swansea, SA1 3SN. https://arcgiveshub.jisc.sc.uk/data/gb216-d/dxch/ddxch/i/hub

Chapter 9
The Nantgarw Porcelain Body

Abstract The composition and formulation of the Nantgarw china body and incorrect recipes given in the literature are discussed against the true composition as revealed by Samuel Walker to Taylor in the 1840s. The role of the components in the high temperature chemistry operating in the kilns is discussed and evaluated against alternative procedures and materials.

Keywords Nantgarw china · Body paste recipes · China formulations · Raw materials · Additives · Soft paste porcelain

At the heart of the historical appreciation of any 18th and 19th Century porcelain factory is the composition of the porcelain paste which comprises the final fired body, which in turn gave the china works its market edge in finish, texture and translucency and from which the varied assortment of useful and decorative wares could be manufactured. Without a doubt, most of the English porcelain factories in the late 18th and early 19th Century made soft paste porcelains, which contained primarily kaolin, bone ash, lime, soda, potash and silica in the form of sand, flint or chert. Special variants appeared which substituted soapstone for one or other of these components—this increased the magnesium silicate content and rendered the fired body much more robust in handling—and also lead (or flint) glass for a greater translucency. The alternative, a hard paste or "true" porcelain mimicked the Chinese imports more closely and is characterised by the absence of calcined bone ash (a phosphatic component) and increased quantities of steatite or soapstone (called *petuntse* in China) (Edwards, *Nantgarw and Swansea Porcelain: An Analytical Perspective* 2018). Generally, hard paste porcelain is colder to the touch and is more robust than soft paste porcelain even when thinly potted. Whereas Derby, Caughley and Spode typify the English soft paste porcelain factories, New Hall and Bristol are examples of hard paste porcelains. Others such as Coalport and Worcester can be properly described in Eccles and Rackham's terminology as "hybrid" porcelains (Eccles and Rackham, *Analysed Specimens of English Porcelain* 1922).

The two Welsh porcelain factories of the early 19th Century, Nantgarw and Swansea, are both described as *exclusively* producing soft paste porcelains: at Swansea, Lewis Dillwyn undertook experiments which started with a glassy porcelain body, progressing to his beautifully translucent duck-egg porcelain, and finally

© Springer Nature Switzerland AG 2019
H. G. M. Edwards, *Porcelain to Silica Bricks*,
https://doi.org/10.1007/978-3-030-10573-0_9

culminating in his attempt to create a more robust *"trident"* body which was commercially unsuccessful, despite its demonstrable ability to withstand cleansing in boiling water (Appendix B). At Nantgarw, William Billingsley finally achieved his ambition to create a highly translucent porcelain comprising mainly kaolin, sand and calcined bone ash, fired at a high temperature—and previous authors have historically stated that he did not attempt to vary this, even when the factory faced economic closure in 1819/20. Whereas Dillwyn's notebooks relating his stepwise experimental and empirical chemical procedures to create a better porcelain paste are still extant (these are reproduced in Eccles and Rackham, *Analysed Specimens of English Porcelain,* 1922; also, here a transcript is given in Appendix B), no such recipes or formulations were recorded by William Billingsley at Nantgarw. After moving to the Coalport China Works at the behest of John Rose from Nantgarw in 1820, William Billingsley and Samuel Walker remained there until the death of Billingsley in 1828. The historical account there also seems to be full of conjecture and several statements made thereafter just do not appear to be verifiable or documented in a forensic sense: for example, we have the following statements which have been generated from the residence of Billingsley and Walker at Coalport:

John Rose employed Billingsley and Walker to manufacture Coalport porcelain from their own Nantgarw formulation and to make porcelain there using the moulds, hardware and kilns which it has been said that he had purchased at the public auctions of surplus Swansea and Nantgarw porcelains held in South Wales in the period 1822–1823.

Following the death of William Billingsley on January 16th, 1828, Samuel Walker remained in Coalport where he *"continued to improve the art of chinamaking at Coalport"* (Jewitt, *Ceramic Art in Great Britain* 1878); he is then reported to have left Coalport and returned briefly to Swansea, where he tried to interest Lewis Dillwyn in re-starting the Swansea factory to manufacture porcelain. We could well enquire here how this would be possible, if in fact, John Rose had already purchased the hardware at the auction sales, as alleged by Jewitt (*Ceramic Art in Great Britain* 1878) and Haslem, (*The Old Derby China Factory* 1876) and removed these items from the site of the Swansea China Works in 1822? It is said that Dillwyn was no longer interested in such a venture and by then his son Lewis Dillwyn Jr. was managing the operations of the Cambrian/Swansea Pottery to manufacture earthenware on the same site. Pottery continued to be manufactured at the Cambrian/Swansea Pottery site until its closure in 1870 and at the adjacent Glamorgan Pottery under the tenure of Baker, Bevans and Irwin between 1813 and 1838.

The subsequent movements of Samuel Walker are now completely unknown, until he resurfaces in the early 1840s, now remarried after the death of his first wife, Sarah, at Swansea, the daughter of William Billingsley, in January 1817, just before Walker and Billingsley had relocated to establish the so-called Phase II of Nantgarw porcelain production some months later in 1817. At this stage, Billingsley had already left Swansea in December 1816, with the intention of re-starting at Nantgarw. In 1847, Taylor published his book, *The Complete Practical Potter*, in which he gives for the first time the composition of the Nantgarw porcelain paste—accredited to a communication to him directly from Samuel Walker. Walker emigrated to the USA, arriving in New York on April 22nd, 1842, on the ship *Monument* from Liverpool with his new wife, also Sarah, where he sets up the Temperance Hill Pottery, West Troy, Albany, New York State (Broderick, *An English Porcelain Maker in West Troy* 1988); Barker, *Pottery and Porcelain in the United States* 1893). He married Sarah, born 1815, with whom he had a daughter Mariah, who was born in 1834. The passenger list gives Samuel Walker aged approximately 50, by profession in his own words: a "mechanic".

Much has been written about the recipe and formulation of the Nantgarw China Works porcelain body and glaze and its revelation to private individuals and also in the public domain, so it is appropriate here to review the historical occurrence of the recipe as proposed by various authors hitherto. It seems that the earliest experiments of William Billingsley in the production of a superior soft paste porcelain can be traced back to his apprentice days at the Derby China Works, where he used a furnace built by his father in the cellar of the family home at Derby, which also involved interaction with his then mentor, Zachariah Boreman (Edwards, *Swansea and Nantgarw Porcelain: A Scientific Reappraisal* 2017): it is conjectural as to whether or not any novel porcelain of worth was created as the business of decorating pieces "in the white" from various factories was at that time quite a viable proposition for ceramic artists. So much so, that Billingsley actually commented on this several times later in his career at Pinxton and at Nantgarw. On leaving Derby in 1795, Billingsley first really started to make porcelain at Pinxton for John Coke and there are accounts of his trials and difficulties encountered in doing this. He was dissatisfied with the Pinxton porcelain body and moved on to Mansfield, where he concentrated upon the decoration of porcelain purchased elsewhere rather than on its primary manufacture. He then moved to Brampton-in-Torksey, where things took a different turn: initially the porcelain manufacturing project there was financed by Henry Bankes, whose financial support ran out in 1805. Four local men stepped into continue the support: William Sharpe, a painter and glazier, Benjamin Booth, a printer, James Walker, a neighbouring farmer, and Samuel Walker, his son. Samuel Walker eventually met Sarah Billingsley and became William Billingsley's son-in-law in 1812. This manufacturing and decorating partnership at Brampton-in-Torksey was formally dissolved on November 21st, 1808; the notice of dissolution was carried in the *London Gazette* on April 5th and 9th, 1808, and also stated that Samuel Walker would assume Billingsley's debts. The importance of Walker's presence at this stage in Billingsley's life cannot be stressed too much as Walker had, it is assumed through contact with William Billingsley, developed a particular skill in furnace design and construction which was to be the foundation of their later process at the Nantgarw China Works. A move to the Royal Worcester China Works of William Billingsley, his two daughters Sarah and Lavinia, and Samuel Walker followed later in 1808, described in a letter from Sarah to her mother dated October 24th, 1808, where Billingsley and Walker at first took on rather menial tasks until they came to the attention of the proprietor, Martin Barr. In 1812, Walker's skill in furnace technology must have impressed Martin Barr, because he was engaged in the construction of a special kiln for porcelain production in the strictest secrecy at Worcester. On September 22nd, Walker married Sarah Billingsley at Worcester and a little while later, in November of the same year, there is a record that

Barr, Flight & Barr accorded the sum of £200 to Billingsley and Walker for services rendered, with the stipulation *"that they hereafter for ever forebear to communicating and imparting the secret of a new method of composing porcelain"*—a transcript of this important letter is given in Appendix A. Martin Barr died on November 10th, 1813, and the new partners, Messrs. Flight, Barr & Barr, in the Royal Worcester Porcelain Factory apparently were not interested in these new experiments and instead set about the production of a cheaper brand of ceramic porcelain. The freedom that had been consequently accorded to Billingsley and Walker there to undertake their clandestine experiments in the formulation and firing of porcelain bodies by Martin Barr must have strengthened their resolve to "go it alone" and shortly after the death of Martin Barr and the negative reaction of Messrs Flight, Barr & Barr to adopting the new formulation they departed Worcester early in 1814. An interesting side-track has been created in the recounting of the experiences of Billingsley and Walker at Worcester, since it is now universally stated that they left seemingly under excellent and amicable terms with the proprietor of the Worcester China Works, Martin Barr, who presented them with a donation of £200 upon their severance and that it is also quite clear that Martin Barr had knowledge of their porcelain making activities at Worcester as he drew up a Bond in which he cautioned them upon payment of a legal penalty of £1000, for any infringement of an agreement not to divulge the formulation of their porcelain recipe to a third person, yet permitting them to manufacture the porcelain themselves (a copy of this Bond is given in Appendix A). This opened the door legally for their manufacture of porcelain at Nantgarw and also immediately argues against previous writers who have suggested that the selection of Nantgarw as their base of operations was purely because it was situated in a very remote part of the country and that the kilns themselves were hidden "away from prying eyes". A possible catalyst for this comment is that when Lewis Dillwyn engaged the services of Billingsley and Walker in 1814 after the failure of the first phase of the Nantgarw China Works operation, he received a very imperious letter from Flight, Barr and Barr warning him against engaging their services to make porcelain otherwise they would take legal action to recover the agreed fixed penalty. This means of course that Dillwyn could never have legally manufactured porcelain of the Nantgarw formulation at Swansea and that Billingsley and Walker could never have revealed the precise recipe to him otherwise they too would be liable for prosecution. So, we would not therefore expect to encounter porcelain made at Swansea of Nantgarw formulation, despite the allegations made by earlier authors who claimed that this must have happened in the earliest days and who have apparently produced some rather diffuse evidence in support of this claim. Another interesting question that can be addressed at this point is why Martin Barr never attempted to engage Billingsley and Walker to make their new porcelain body at Worcester: after all, they were his registered employees when they conducted their experiments there and any refusal on their part would surely not have been met with a generous "parting gift" of £200? It can be surmised that the decoration of the Barr, Flight and Barr porcelain body then being made at Worcester was perfectly fitting to their clientele and that they did not wish to interrupt production schedules and change well-established kiln-firing characteristics to compensate for an unknown and untested quantity. The timing of Martin Barr's gift to Billingsley and

Walker and his death in 1813 are very close and the refusal to proceed further with the new ideas in porcelain manufacture adopted by Messrs. Flight, Barr and Barr must have come as a shock to them, hastening their departure from Worcester in early 1814.

This historical and chronological discussion sets the scene for a detailed assessment to be made of the claims made by subsequent historians, authors and chroniclers relating to—who knew the recipe and what happened to it? The discussion is compromised in several instances because clearly some formulations are quoted which bear no resemblance at all to a Nantgarw paste and these can be dismissed now as pure fantasy or wishful thinking!

9.1 The Nantgarw Porcelain Body Recipe

John Taylor: in "The Complete Practical Potter" (1847).
Taylor quotes a recipe for the Nantgarw body apparently provided by Samuel Walker, namely, 26 lbs bone, 14 lbs Lynn sand and 2 lbs potash, mixed with water, made into bricks, fired in the hard part of a biscuit kiln then pounded and ground for use. Of this frit, 40 lbs were mixed with 20 lbs of china clay and made into the body paste. Professor J. W. Mellor in the 1880s made up two sets from this recipe, which were both fired at Minton's China Works in their biscuit oven. Casting was employed with both sets of paste, for which the body seemed rather non-plastic but yet fired well. Professor Mellor thought that the ageing of the paste on storage as a wet slip would improve the plasticity—and he believed that this would have been the case at the Nantgarw China Works, where local documents refer to a "sleep-house" or slip room for such storage to take place. Professor Mellor also suggested that kneading and treading of the clay frit would have seen a significant increase in plasticity and homogeneity of the paste.

C. F. Binns: in "The Potters Craft: A Practical Guide for the Studio and Workshop" (1922, p. 27).
The Nantgarw china body was a soft paste comprising a frit of 210 lbs sand and 30 lbs of pearl ash: the body consisted of a mixture of 150 lbs frit, 220 lbs bone ash and 65 lbs of china clay. Despite proposing a recipe that contained almost 50% bone ash, Professor Mellor deduced that this composition was actually totally ineffective as it had too little china clay for plasticity.

John Haslem: in "The Old Derby China Factory", (1876, p. 246).
Haslem gave the following recipe for the Nantgarw body: "*Bone* 5 lbs, *Lynn sand* 5 lbs, *potash* 5 lbs, *mixed, made into bricks and calcined in the biscuit kiln. The resulting frit is then ground and mixed with Cornish clay to suit your mind*". Haslem stated that this recipe was given personally by Billingsley to George Hancock at Worcester (and cited as such by Turner in 1897 in *The Ceramics of Swansea and Nantgarw*). It should be noted that this recipe is totally devoid of bone ash and Professor Mellor has noted that it is almost impossible to work this formulaic paste even with a proportion of Cornish clay approaching 20%. This body is unsuitable

for moulding and Professor Mellor reckons that it is "*nothing like Nantgarw*". The comment must be made here that there would be no reason for Billingsley to reveal his formulation to Hancock and certainly none for his providing an incorrect one in that it was missing an important and vital ingredient, namely calcined bone ash—which Billingsley believed was his secret component within a secret recipe formulation.

Anonymous: *The Pottery Gazette, December 1st* (1885).
An anonymous contributor reported a very similar body composition to that reported by John Taylor in his book of 1847 (*The Complete Practical Potter*). This comprised 13 parts of bone, 7 parts sand and 1 part of potash, sifted, mixed well together and fired in a strong biscuit heat. The procedure was completed by the addition of 20 parts of china clay to 40 parts of the above frit, grinding them well and followed by drying in a slip kiln. This recipe was apparently given to the anonymous contributor who was born in Nantgarw and still lived there after the closure of the Nantgarw China Works by Samuel Walker personally. Another variant of this recipe, which had the addition of 100 parts soapstone, coarsely ground, and 150 parts china clay, was that of Walker's "*French body*".

Professor J. W. Mellor: 1885
The distinguished ceramic chemist and analyst, Professor J. W. Mellor, at the instigation of the Nantgarw historian and researcher, Ernest Morton Nance (*The Pottery and Porcelain of Swansea and Nantgarw* 1942), undertook the synthesis of Nantgarw porcelain in 1885, aided by his two students, Messrs. Lignan and Middleton, using the recipe provided by Taylor in *The Complete Practical Potter* in (1847), apparently provided by Samuel Walker. The outcome of the series of experiments carried out by Professor Mellor was that a porcelain indistinguishable from Nantgarw was successfully made and glazed at these trials. It is interesting too that Morton Nance cites a note held in the Coalport China Works archive giving details of the Nantgarw paste which is identical to the recipe provided by Taylor! What cannot be deduced, however, is whether or not this appeared after the publication of Taylor's book in (1847) or if Billingsley and/or Walker had in fact provided the recipe to John Rose before Billingsley died in 1828 or before Walker later left Coalport. Nevertheless, Professor Mellor's experiments in recreating Nantgarw porcelain using a kiln firing temperature of 1300 °C were judged eminently successful according to contemporary independent experts and observers such as Ernest Morton Nance.

William Weston Young, 1821.
Turner (*The Ceramics of Swansea and Nantgarw* 1897) reports a formula for the Nantgarw porcelain recipe recorded in William Weston Young's diaries as follows: 4 parts sand, 1 part soaprock, ½ alkali for the frit, of which 5 parts are then mixed with 1 part of china clay: it should be noted that this body is completely devoid of the vital component of bone ash as has been remarked upon earlier and only resembles Swansea porcelain superficially in the presence of the soaprock, but again misses out the addition of flint glass cullet specifically occurring for several of Dillwyn's formulations. Other recipes cited by Young include four variants given as follows:

No. 1: 8 parts sand, 4 parts soaprock and 1 part potash;
No. 2: 8 parts sand, 2 parts soaprock and 1 part potash;
No. 3: 8 parts sand, 2 parts soaprock and 1½ parts potash;
No. 4: 8 parts sand, 1 part soaprock and 1½ parts potash;
For each of these frits, 5 parts of frit are mixed with 1 part china clay to form the porcelain paste. As a comparative example, the "true" Billingsley Nantgarw recipe can be given as:
7 parts sand, 13 parts bone ash and 1 part potash, as a frit of which 5 parts are mixed with 2½ parts of china clay—comparison thus indicates a significantly very different composition indeed for the body paste! A summary of the porcelain trials carried out by William Weston Young and noted in his diary is given in Table 4.1.

Simeon Shaw, in "Chemistry of the Heterogeneous Compounds", 1837.
"Mr Billingsley at the Nantgarrow factory from Lynn sand, potash and other compounds made a porcelain which as an artificial feldspar had some excellence".
This is a curious statement as Nantgarw porcelain is nothing like feldspar and its major components have been totally ignored, yet Shaw has correctly identified the source of the pure white sand used by Billingsley. In fairness to Shaw, the chemical analysis of Nantgarw or any other porcelains were not available at this time and the Nantgarw formulation recipe was also not in the public domain, so much of the statement must be based on dubious hearsay and assumption. Clearly, there was no suggestion that the Nantgarw recipe was circulating publically at this time and all was still "under wraps" and a still closely guarded secret.

It is relevant at this point to examine closely the recipe of William Weston Young cited by Turner (item number 6 above and given in *The Ceramics of Swansea and Nantgarw* 1897), which states that the composition of Nantgarw porcelain comprised only four components, namely, sand, soaprock, china clay and potash/alkali—this, of course, is more akin to hard paste china than the characteristic soft paste of the true Nantgarw porcelain. However, when we actually look at the relative component composition as given by Turner and compare this with those quoted in William Weston Young's Diaries then we have rather more strange anomalies. Firstly, Turner in his book (*The Ceramics of Swansea and Nantgarw* 1897) has given the actual component raw materials and note those cited in code by Young (here reproduced in Table 4.1) with no indication or justification given for this translation. Comparison with the actual compositions given by Young in his diary for October 1819 with the data given by Turner reveals some further peculiar discrepancies: Turner gives the Young formulation as 4 parts sand, 1 part soaprock and ½ part of alkali forming a frit, of which 5 parts were then mixed with 1 part china clay. Here, in Table 4.1, without the provision of a key it is not possible to definitively transcribe Young's codes into the materials used, but it is immediately apparent that Turner has made a simplification—only four components are assumed by Turner to have been used by Young, namely, china clay, sand, alkali and soaprock, whereas Young cites 8 distinct substance codes, namely, FO, V, NC, B, HC, K, L, and N (Table 4.1). The present author has attempted to ascribe an identification to each of these from their potential compositional presence and using a cross-correlation with the codes favoured by

Lewis Dillwyn and given in his own notebook (Appendix B; Eccles and Rackham 1922); five of these codes are found to match the raw materials in these two recipes. Secondly, the compositional ratios of materials cited by Turner for Young's recipes do not equate with the data given in the diary notes shown in Table 4.1. For example, Turner gives Young's main formulation as 4 parts sand, 1 part soaprock and ½ part alkali—5 parts of which are then mixed with 1 part of china clay, whereas Young gives the formulation of his No. 1 frit as 4 parts sand, 1 and ½ parts china clay and ¼ part alkali, of which 2 parts are fritted and ground with 1 part china clay. Also, Turner cites just four more variations in composition made by Young when in fact there are nine more given in Young's handwritten notes: a significant difference lies in the fact that Young only uses soaprock in the final three of these, namely Nos. 8, 9 and 10, whereas Turner has him using this material as the major component in all five of his formulations represented there. Turner has ignored any contribution to the paste frit from nitre, borax, lime, lead glass and ball clay. Something is amiss here and whereas Turner could be forgiven for not appreciating the importance of the minor materials in Young's formulation, for which Dillwyn's recipes would have helped (although it should be pointed out that these did not surface until 1922) when they were lodged in the V&A Museum archives), his failure to record Young's formulation changes over ten recipes is rather incongruous.

In summary, therefore, it is apparent that William Weston Young's trial porcelain formulations which he tried during 1820–1821 at Nantgarw are very significantly different from those which constituted the true Nantgarw soft paste porcelain as made by Billingsley and Walker from 1817–1819. The major conclusion that can be noted from Young's experiments is that he never appreciated the use of bone ash in Billingsley and Walker's Nantgarw formulation—the secret within secrets had been very successfully preserved!

At this point, it is relevant to reiterate the Billingsley/Walker recipe for their Nantgarw porcelain body, which comprises ground bone ash, quartz sand and a potash flux which is fired at a high temperature into a glassy matrix which is then ground finely and mixed with china clay and water to produce the final paste. Maurice Hillis (Hillis, *Welsh Ceramics in Context*, Volume II 2005) has correctly pointed out that the fritting part of this initial process is extremely important since this renders insoluble any component that would dissolve out of the mixture when the addition of water is made to compose the paste. Also, any gas-forming materials at the elevated temperatures employed in the fritting process would be decomposed to facilitate the escape of gas bubbles which otherwise could seriously compromise the translucency of the biscuit porcelain in the firing kiln by entrapment in the solidifying matrix. On these two counts, therefore, the potash used as a flux in the first formulation to form the frit would be leached out in water and thereby lost, and any calcite added along with or in lieu of lime would be decomposed at 700 °C to calcium oxide and so releasing carbon dioxide bubbles into the porcelain paste. Dillwyn, as a practising chemist, is on record as saying that he deplored the substitution of calcite or alabaster instead of lime in porcelain bodies for this very reason as the formation of gas bubbles in the resultant porcelain body was detrimental to its translucency.

9.2   The Nantgarw Glazes

William Weston Young, 1821.
These were naturally used after the departure of Billingsley and Walker from Nantgarw in 1820: Young's diaries reveal the following compositions for two glazes that
he tried at Nantgarw. The first of these, his No. 2 glaze, comprised simply 5 parts of
Lynn sand, 4 parts of borax as a frit to which was added 1 part of lead oxide and 1 part
of glass. Young's No. 2 glaze made a small variation on the No. 1 glaze, in that he
increased the ratio of lead oxide to glass to 2 parts of lead oxide to 1 part of glass for
the addition to the frit. It was clear from his diaries that Young tried both glazes, but
he does not specify a preference. Professor Mellor tried both Young glazes and found
them to be satisfactory, but they had a tendency to craze, and he also commented
that they were less satisfactory than the Billingsley glaze which was smoother to
work and had the better appearance after firing. Young also commented on a curious
pigskin like effect which he noted on Pardoe's work with his glaze; Professor Mellor
attributed this to the glass composition used in the formulation, which he fired at
between 1100 and 1120 °C (his experimental thermal cones numbered, 1a and 2a)
and that the use of more recent flint glass proved to be successful.

William Billingsley, 1817.
The formulation for Billingsley's Nantagrw glaze, a very white glaze, was never
stated by Billingsley himself or by Samuel Walker, but details are given in John
Taylor's book (see below), which apparently are both not perfectly compatible with
the biscuit china as concluded by Professor Mellor and his colleagues in their later
experiments conducted in 1885.

John Taylor: in "The Complete Practical Potter" (1847).
Taylor wrote that there were two Nantgarw glazes used in the lifetime of the factory:
the first of these, the No. 1 glaze comprised 14 lbs of china clay, 18 lbs of sand, 14 lbs
of bone, 13 ½ lbs feldspar, 12 ½ lbs china stone, 11 ½ lbs flint and 110 lbs borax, to
which was then added 110 lbs lead oxide.

 The No. 2 glaze, comprised 8 lbs sand, 6 lbs chalk, 2 ½ lbs china stone, 40 lbs china
clay, 3 ½ lbs feldspar, ½ lb flint, 6 lbs soda and 3 lbs nitre, all of which were fritted
together. Interestingly, Taylor confirms that Billingsley only used the No. 1 glaze
and he does not ascribe the derivation of the glaze compositions to Samuel Walker,
unlike the body paste. The No. 2 glaze, a leadless glaze, he states was adopted by
Nantgarw prior to the sale of the factory in 1820 and the departure of Billingsley and
Walker to Coalport. Professor Mellor was negative about the effectiveness of both
glazes; he claimed that No. 1 was not satisfactory and that No. 2 just did not work at
all as it had a different melting range to that required for the body.

Professor J. W. Mellor: 1885.
This provides perhaps the best and earliest example of experiments designed te
emulate the Nantgarw china paste and glaze compositions after firing. Details are
given above.

William Morgan, 1926.

Corresponded with Ernest Morton Nance for his book *The Pottery and Porcelain of (Swansea and Nantgarw* 1942); he was born in Nantgarw and obtained much information from *"old folks still alive in 1875"* who had actually worked at the Nantgarw China Works. He wrote an essay in October 1875, entitled *"The Nantgarw Ancient China Works"* for a local eisteddfod at Tongwynlais, citing himself as William Morgan of Taff's Well. Turner attempted unsuccessfully to reproduce this essay in his book on *The Ceramics of Swansea and Nantgarw* published in 1897, but eventually he had to rely on Robert Drane's notes. Morton Nance (*The Pottery and Porcelain of Swansea and Nantgarw*, 1942, p. 364) gives the Nantgarw glaze recipe he obtained from William Morgan as follows:

50 parts sand, 60 parts borax, 20 parts whiting, 4 parts nitre and 4 parts lead oxide fired in a glost oven, then finely ground and 50 lbs of the frit was mixed with 30 lbs of china stone, 4 lbs of china clay, 4 lbs of whiting and 4 lbs of lead oxide. The glaze was applied to the biscuit porcelain, which had itself been fired between 1270 and 1280° C, and re-fired between 100 and 1120 °C in a glost kiln. Morgan also remarks that William Billingsley had a secret and private workroom in the basement of his house where he added personally bone ash and clay to the frits—this would, of course, have preserved the secrecy regarding the quantitation of the Nantgarw glaze and paste body formulation.

The Pottery Gazette, 1885.

A glaze composition was cited as follows: 50 parts sand, 60 parts borax, 20 parts whiting, 4 parts nitre and 4 parts lead oxide were mixed and fired then to this frit was added 30 parts china stone, 4 parts china clay, 4 parts whiting and 4 parts lead oxide. It was suggested that the origin of this glaze came from Samuel Walker and Professor Mellor approved its effectiveness—it was a successful glaze and had the *"correct composition in every respect"*.

9.3   Chemical Components of the Raw Materials in Nantgarw Paste Formulations

It is appropriate here to list the individual mineral components of porcelain manufacture as adopted by the Welsh factories, particularly Nantgarw, as operated firstly under the stewardship of Billingsley and Walker and then under Young and Pardoe. These components will dictate the durability and eventual translucency of the finished article after firing at high temperatures as has been elaborated elsewhere in greater detail (Edwards, *Nantgarw and Swansea Porcelain: An Analytical Perspective* 2018).

Sand: the finest quartz river sand was specified by many china works business directors to provide the quartz silica, SiO_2, necessary for preparation of the basic porcelain paste matrix. Sand comprises generically finely-divided rock and mineral particles, its geological descriptor occupying the particulate size range between gravel and silt. Fine river sand, or white quartz sand, has a particle size range broadly between 0.06 and 0.1 mm which is often coloured by the presence of transition metal oxides, such as iron (III) oxide, Fe_2O_3, and infusible particulate matter such as carbon or manganese (IV) oxide, pyrolusite, MnO_2. Although a pure white sand is found which can also contain gypsum, sand derived from the weathering of basalts can display a range of colours from green to brown and even black in volcanic regions, due to the presence of minerals such as olivine and magnetite and these can generate blemishes in the glassy matrix formed upon fusion in the firing kilns, with a resulting deleterious effect on the porcelain translucency and in some cases rendering the final article unsaleable. When occurring near the surface of a porcelain item, however created, these blemishes often cause a pitting or grittiness in texture which is not acceptable to the clientele: in several cases, therefore, ceramic artists have disguised such surface blemishes in an otherwise fine piece of porcelain by applying an enamelled insect such as a moth or a butterfly, or perhaps a single flower bud, which masks the small area affected—examples of this can be seen on the underside of the rim of a very large Nantgarw platter from the *Farnley Hall* service (Fig. 9.1) which has a trio of mosquito-like insects masking a surface blemish (Fig. 9.2). This prestigious service was commissioned by William Ramsden Hawkesworth Fawkes, an influential London businessman, Yorkshire landowner and MP, probably from John Mortlock's in Oxford Street, London, and would have been decorated in the Robbins and Randall atelier (Edwards, *Swansea and Nantgarw Porcelains: A Scientific Reappraisal* 2017): clearly, the acceptance of such a very large and otherwise perfect piece of Nantgarw porcelain sent out from the factory was not problematic for the purchasers "in the white" for subsequent decoration in London and the addition of little insect motifs on a defective under-surface area that was not immediately visible from above was deemed to be quite acceptable. It is realised, therefore, that porcelain manufacturers such as Dillwyn and Billingsley would have prioritises their selection of a superior quality of fine quartz sand as being critically important for the realisation of their objective of producing a very high-quality and translucent porcelain: in this respect, the sourcing by William Billingsley of Lynn sand from the Norfolk gravel pits can be appreciated, despite the extensive additional transportation cost involved over that from comparative local river sand.

Smalt: a cobalt blue glass (not to be confused with the mineral smaltite, a cobalt arsenide) which is used finely ground in the glassy matrix to combat or offset any background residual yellow or brown colouration which may occur from impurities in the matrix, especially iron oxides from the sand. It is also a component of the glaze slip into which the fired porcelain piece is dipped for a sealing coat of a high gloss, transparent white finish. Cobalt blue glass consists of cobalt (III) oxide, Co_2O_3, added to an aluminosilicate glass, typically comprising SiO_2 65%, K_2O 15%, Al_2O_3 5% and Co_2O_3 10%. Frequently represented incorrectly as $CoO.SiO_2$, and also alternatively and incorrectly as a cobalt aluminate, cobalt blue is really a cobalt aluminosilicate and has been known from about 1500 AD. It has often been confused with the more ancient historic *Egyptian blue*, which is a calcium copper (II) silicate, $CaCuSi_4O_{10}$, also termed *cuprorivaite*. *Egyptian blue* was synthesised in antiquity, it is believed around 3500 BC. Cobalt blue was used extensively in blue transfer earthenware patterns in the 18th and 19th Centuries and the famous "*Bristol blue*" coloured glass

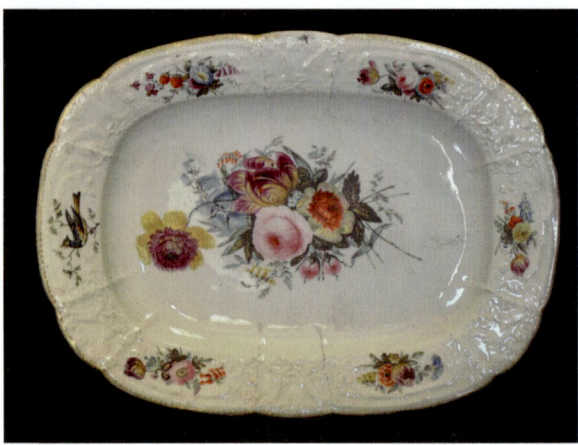

Fig. 9.1 Large rectangular meat platter from the *Farnley Hall* service of Nantgarw porcelain, unmarked except for an impressed 4 and 7, superbly decorated in London with dentil edge gilding in the manner of the generic *Brace* service, with fruit, flowers and birds in vignettes and typical Nantgarw moulding at the edge. Located and identified by the author in the cellars of Farnley Hall, Otley, North Yorkshire, in March 2016—the remains of this once extensive service of over a hundred pieces which was probably purchased from Mortlock's by Walter Ramsden Hawkesworth Fawkes MP in 1817–1819, now amount to approximately 30 pieces, including some very substantial and large platters. Most plates, soup dishes and tureen stands belonging to this service are marked with the impressed mark NANT-GARW C.W. A frequent visitor to Farnley Hall at this time was J.M.W. Turner, who spent many months painting there with his friend Walter Fawkes, and it is tempting to suggest that the two men may have dined off this superb porcelain dinner-dessert service. Reproduced with permission of Guy Fawkes Esq., Farnley Hall, Otley, North Yorkshire

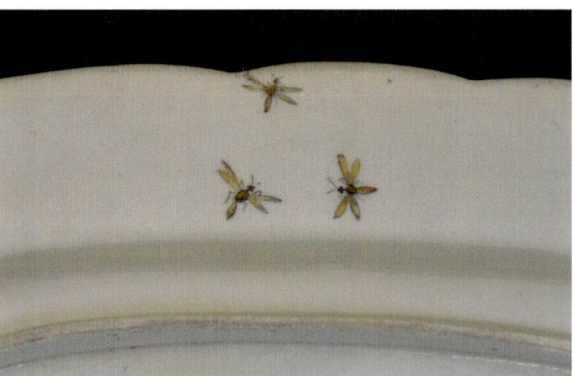

Fig. 9.2 Reverse of large meat platter shown in Fig. 9.1 from the *Farnley Hall* service of Nantgarw porcelain, unmarked except for an impressed 4 and 7, superbly decorated in London with dentil edge gilding, showing three mosquitoes strategically placed to cover surface blemishes. Reproduced with permission of Guy Fawkes Esq., Farnley Hall, Otley, North Yorkshire

of bottles and drinking vessels of the Georgian and Regency periods reflect the trade in this valuable coloured oxide through the port of Bristol. Dillwyn certainly recorded his use of smalt in his finest duck-egg Swansea porcelain but is not clear if Billingsley adopted this also in Nantgarw: it is suspected that he possibly did not, as smalt did not feature in the recipe details provided to Taylor (*The Complete Potter* 1847) some years later by Samuel Walker.

Borax: is sodium tetraborate decahydrate, $Na_2B_4O_7.10H_2O$, and is used as a flux to aid the fusion of the glassy components of the paste in the firing kiln—this meant that the paste would become more fusible at a lower temperature and this would assist in the better preservation of the articles in the kiln.

Alum: alum is a hydrated aluminium potassium sulfate, $KAl(SO_4)_2.12H_2O$, and has been described as an unusual additive to porcelain bodies (Ramsay and Ramsay 2007, 2017). Church (*English Porcelain* 1894) has stated that the paste of *Vieux Sevres*, also known as *pate tendre,* comprises 8 parts marl, 17 parts chalk, and 75 parts glass frit—the latter component containing 3 parts per 100 of alum. Ramsay and Ramsay (2007) draw attention to this unusual additive and postulate the origin of this idea, but they also note that early Bow porcelains have been found to contain alum as a minor component. Clearly, the origin of the sulfur presence in derived analytical data can be ascribed variously to gypsum, anhydrite (dehydrated gypsum), pyrites or alum and the analytical presence of sulfur is not in itself an indicator of an alum additive.

Clay: clay is a generic name for the geologically weathered materials derived from feldspar, granites and basalts. It covers many types of structure based on three-dimensional silicate matrices with interstitial metal ions and water molecules, several of which have Si-O-Si bridging bonds which renders different degrees of pliability and hardness upon the silicate skeletal structures. *Kaolinite* and *china clay* are primary or residual clays formed geologically from granite basins as distinct from *ball clay* (also known as pipe clay, blue clay and plastic clay), which is a secondary or transported detrital clay found in sedimentary basins. The selection of the type of clay utilised in porcelain manufacture is critically important for plasticity and fusion properties at high firing temperatures in the kilns. Generally, two types of clay were used in Welsh porcelains, namely *ball clay* and *kaolin* (*china clay*), the latter being the major clay component in any paste mixture. *Ball clay* is an extremely fine kaolinic sedimentary material with a low haematite content, so it is very white in colour, whose inclusion in a porcelain paste before firing renders a greater plasticity, a better rheology and ease of working. It has a composition approximating within a broad range to 20–80% kaolinite, 10–25% mica and 6–65% quartz and may also contain carbonaceous material such as lignite, which is burned off upon calcination. It derives its name from the mining operations which provide this clay in cubes weighing between 15 and 17 kg with the corners rounded off. It is found in Eocene deposits (40–50 Mya) in SW England and especially in Cornwall and Devon. It is believed that this material comprises the "*virgin earth*" mentioned by Eccles and Rackham (*Analysed Specimens of English Porcelain*, 1922) and Church (*English Porcelain*, 1894): its expected contaminants are siderite, pyrites, anatase, gypsum and dolomite. On the other hand, *kaolin (china clay), also* called kaolinite, and formulated as $Al_2Si_2O_5(OH)_4$, is a layered aluminosilicate of the phyllosilicate type and a member of the serpentine group, structurally containing tetrahedral sheets of SiO_4 units linked to sheets of AlO_6 octahedra and this is produced geologically by the weathering of feldspar. The name *kaolin* is derived from a corruption of the Chinese Kauling (alt. *Gaoling*), the name of a hill near Jauchau Fu, where the material was first mined to produce hard paste porcelain (Dana, *Textbook of Mineralogy* 1955) near to the first porcelain kilns set up at Jingdzehen in Jiangxi province. It is sometimes represented as $Al_2O_3.2SiO_2.2H_2O$, which can assist in the interpretation of the wet chemical analyses of porcelain through the quantitative estimation of the silica and alumina contents. It is appropriate here to describe the complex changes which occur in the structure of *kaolin* when subjected to the heat of a ceramics firing kiln: firstly, dehydration (loss of molecular water) commences at 550–600 °C to form *metakaolinite* with loss of further hydroxyl (OH) groups up to 900 °C. At this temperature, the first chemical skeletal structural change occurs in the silicate matrix, when the original $Al_2Si_2O_5(OH)_4$ undergoes transformation to $Al_2Si_2O_7$ with the formal loss of two molecules of water. At the slightly higher temperature range of 925–950 °C a spinel is formed when two molecules of the dehydrated aluminosilicate form $Si_3Al_4O_{12}$, with the elimination of a molecule of SiO_2. Around 1050 °C three molecules of this spinel now react to form mullite ($3Al_2O_3.2SiO_2$) with the elimination of 5 molecules of silica as SiO_2 in its high temperature form, *cristobalite*. Finally, at the very highest kiln temperatures of 1400 °C, stability is achieved with some internal rearrangements being made to the needle-like structure of mullite. Ramsay and Ramsay (2007, 2017) have defined five different types of clay from various sources used at Bow for porcelain manufacture with the following compositional ranges: silica 44.8–62.69%, alumina 0.49–38.4%, magnesia 0.03–29.93% and lime 0.06–0.3%—all of which contain haematite at about 1%. In addition, several recipes at Bow, at Swansea and Nantgarw and at other factories involve a mixture of *kaolin* and *ball clay* obtained from different Cornish mines: the typical composition of *kaolin* is given as silica 44.8%, alumina 38.4%, magnesia 0.03%, potash 3.33 and soda 0.06%, whereas similar values for the *ball clay* are 54, 30.0, 0.51, 3.1 and 0.46%, respectively. Unsubscribed mixtures of these two clays will therefore defer the analytical interpretations even further to that expected for compositional differences between the alternative mine sources and reinforces points made by earlier authors about the source-dependent variation in composition of the raw materials used throughout porcelain manufactory production.

Soaprock or *steatite*: a metamorphic schist silicate which is rich in magnesium—a variant of talc, often formulated as $3MgO.4SiO_2.H_2O$ or $H_2Mg_3(SiO_3)_4$, it is sometimes described simply as an acid magnesium silicate. A talc formed from the metamorphism of ultramafic protoliths such as serpentinite, a typical steatite is comprised of 63.4% SiO_2, 31.9% MgO and 4.7% H_2O), with the minor occurrence of CaO and Al_2O_3 (Anthony et al. *Handbook of Mineralogy, Volume VII: Silica and Silicates* 1995). This material has often been identified and correlated directly and incorrectly as it transpires with the *petuntse* (sometimes referred to as *porcelain stone* or *china stone*) of the Chinese hard paste porcelains where it confers a robustness upon the fired porcelain body; it is frequently found in admixture with *kaolin*, but the texture and translucency of the body can potentially be compromised. Nevertheless, Dillwyn used such a mixture in his later attempts to improve the strength of Swansea china, but the resultant *"trident"* paste was not appreciated by clients, falling short of the fine quality of the duck-egg paste it replaced despite the excellent floral art displayed in the enamelling of these pieces. Soapstone generally contains about 1/3rd of its weight of magnesia, MgO, so analysts have multiplied the magnesia value by 3X to obtain the soapstone content of a particular porcelain item. The soapstone content of some porcelains is significant, for example, a Worcester body formulated by Martin Barr in 1800 gives the following recipe—Lynn sand 300 parts and flint glass 15 parts are calcined in a biscuit oven; then of this frit, 300 parts are mixed with 240 parts soapstone for the final porcelain body. This, therefore, represents a soapstone composition of 48%—yet analytical figures on pieces from this same period give a magnesia content of 10.5%, which on the basis of the above conversion factor, represents a soapstone content of only 31.5%. Clearly, this discrepancy gives cause for concern, as there is a 30% error between the determined analytical composition and the established recipe from the factory work books which requires explanation and we need to examine more closely the incipient errors made in the weighing practices of each component, variance in sourcing of the raw materials and mechanical deviations in the frit preparation. It has already been stated that the corresponding Swansea *"trident"* body formulation accepted finally by Dillwyn after his experiments comprises: *"Body No. 2, silica 4 parts, soapstone 1 part and potash ½ part"* (Appendix B), so the precision of the weighing and preparation of the paste mixture is seemingly rather vague and open to incipient errors in formulation although on the surface it appears that the trident body soapstone content is only about 19%, considerably less than the Worcester version of Barr, Flight & Barr cited above. Despite this, Dillwyn's departure from a high kaolin soft paste porcelain which proved to be so successful, commercially marketed and appreciated as his famous duck-egg formulation, into the less well appreciated but more robust *trident* variation containing soapstone proved a commercial disaster, yet the Worcester factory was still able to maintain a successful market edge in this area, despite the significantly higher levels of soapstone used in the porcelain paste. At this point it is relevant to indicate a potential source of confusion which can arise from the rather incautious usage of the three terms *soapstone, china stone* and *soaprock*: strictly, these are very different mineralogically. *China stone* is a rare, medium-grained, feldspar-rich partially kaolinized derivative of granite which is characterised by the absence of iron-bearing minerals and contains quartz, mica, feldspar, kaolinite and traces of fluorspar. This mineral occurs only the neighbourhood of St. Austell in Cornwall in the UK and should not be identified strictly with *petuntse* (*baidunzi*), the Chinese kaolin or porcelain stone, technically incorrectly termed "china stone" which is also a micaceous and feldspathic material used in the production of Chinese hard paste porcelain. *Soaprock* is different again in that it is a purer form of steatite, a talc schist and chemically described as a hydrated magnesium silicate. These terms are frequently confused in the literature, although the chemically trained Lewis Dillwyn does identify them separately and correctly in his recipes and formulations as FO (representing *china stone*, or alternatively *composition*) and SR (representing *soaprock*). *Petuntse* on the other hand is often mis-described as *china stone* because of its Chinese origins and is more correctly described as porcelain stone as it is of a different mineralogy to *soaprock, china stone* and *soapstone.* Hence, the former (*petuntse*) can be reserved exclusively for the Chinese hard paste porcelains whereas the latter variants can be regarded as components of English and Welsh soft paste and hard paste porcelains. Soapstone is also known as soaprock or steatite. A metamorphic talc-schist which is rich in magnesium, it is mainly talc with small amounts of chlorite and amphiboles (tremolite, cummingtonite and anthophyllite). It is formed geologically by the metamorphism of ultramafic protoliths (such as dunite and serpentine) and the metasomatism of silicious dolostones. Soapstone is sometimes termed pyrophyllite: it is relatively rather soft, with a hardness of 1 on the Moh's scale of hardness. In contrast, for steatite, the amount of talc content varies greatly—30% comprises a hard steatite, whereas 80% is a soft steatite. When heated to 1000–1200 °C it is converted to enstatite and cristobalite. Talc, a hydrated magnesium silicate, has the chemical formulation (Dana, *Textbook of Mineralogy* 1955), $Mg_3Si_4O_{10}(OH)_2$.

Pearl ash: this component has also caused some confusion historically in attempts to describe its chemical composition. It is technically potassium carbonate, K_2CO_3, and is usually sourced naturally in carbonaceous deposits along with its calcium analogues—however, it is mined as an impure material and since the isolation of pure potassium carbonate in industrial quantities from the raw material is rather expensive and quite prohibitive for commercial porcelain manufacture, it was used as a less pure yet refined form of "potash", which gave rise to its estimated presence in analytical determinations simply as K_2O. It was normally baked in a kiln to remove volatile impurities before use.

Alabaster: although chemically this is a translucent form of hydrated calcium sulfate, gypsum, $CaSO_4.2H_2O$, in earlier times this terminology was also applied to translucent forms of calcium carbonate called *onyx marble* (where it is more correctly termed geologically as calcite or aragonite, two forms of naturally occurring calcium carbonate, $CaCO_3$); it is not therefore to be confused with onyx, which is a coloured silicate related to jadeite, a metasilicate of sodium and aluminium formulated as $NaAl(SiO_3)_2$ or alternatively $Na_2O.Al_2O_3.4SiO_2$. Although Josiah Wedgwood was an adherent of the use of gypsum and alabaster in the manufacture of his creamy Queen's ware from 1759, Lewis Dillwyn is on record as saying that the presence of alabaster could be detrimental to the Swansea porcelain body because of its "blistering" effect—this could be ascribed chemically to its decomposition around 650–700 °C, releasing gaseous carbon dioxide and forming lime, calcium oxide, CaO. The lime would not in itself be a problem as this was often added as a constituent in the porcelain paste mix before firing anyway and would be expected to react with the acid silicates in the mixture to form calcium silicates, but the formation of CO_2 bubbles in the paste reactants in the kiln at elevated temperatures would certainly have created voids in the plastic body and give rise to a lumpiness and blistering in the viscous paste. It can be assumed, therefore, that Dillwyn was referring to the calcite or aragonite connotation for alabaster as gypsum would not decompose in the same way in the kiln, merely dehydrating to anhydrite, $CaSO_4$, and thereafter remaining stable. In the latter instance, water molecules are thermally removed from the gypsum at a temperature between 170 and 250 °C; it is reasonable to assume that if Dillwyn was in fact referring to the gypsum connotation for his reference to alabaster then the removal of water at this low temperature would not be viewed as having such a detrimental effect upon the paste undergoing firing as this would still have been moist anyway through the retention of some water at this temperature. Hence, we can infer that Dillwyn was warning about the effect of carbonate decomposition at more elevated temperatures, when the effect of carbon dioxide emission upon the then drier porcelain body would have been more unsustainable in the kiln firing process.

Flint and flint glass: these are chemically similar in that the major component is silica, SiO_2, and indeed historically they both derive from the same root in that flint nodules from South East England were first used as a source of high purity silica by George Ravenscroft in the 1670 s to make his renowned heavy potash lead glass of high refractivity and dispersion, which contained nominally 20–60% lead oxide. Whereas flint occurs as glassy nodules in chalk deposits and can be coloured grey to brownish red by metal oxide impurities, especially iron III) compounds, it is relatively easy to select manually the whitish-blue and grey nodules which represent a very high purity silica content with minimal impurities present. Flint is actually a higher quality form of *chert*, which is a fine-grained sedimentary rock containing microcrystalline silica occurring in limestone, chalk and marls: other special forms of chert are chalcedony, jadeite and agates. The main difference, therefore, between flint and flint glass is the significant lead content of the latter, which can be detected by chemical analysis, bearing in mind that lead signals can also be observed from the superficial glaze applied to finished porcelains. Essentially, therefore, we must recognise the chemical distinction which appears in recipes which indicate the use of ground frits containing flints and/or flint glass in porcelain pastes. Reference to Lewis Dillwyn's recipes for his Swansea duck-egg porcelain reveals that these involved the addition of flint glass to his frit, whereas the recipe purporting to be from Nantgarw given to Taylor in 1847 (*The Complete Practical Potter* 1847) by Samuel Walker specifically excludes this additive. Hence, a very positive analytical discrimination between Swansea and Nantgarw porcelains would be expected to be the analytical marker for lead in the former which would be absent in the latter. Unfortunately, this seems to have escaped attention until now and none of the analytical work carried out to date mentions the importance of the determination of lead content in Swansea or Nantgarw porcelains: the silica content determinations do not assist here since both flints and flint glass are high silica content materials and, therefore, both Swansea and Nantgarw porcelains have a major component presence attributable directly to silica.

Bone ash: this is undoubtedly one of the most important component raw materials of Nantgarw porcelain and paradoxically the least well understood. Its preparation was afforded special attention by Billingsley and the mixing of the Nantgarw paste raw material components was accomplished in total secrecy and personally under-taken by him in a basement room under lock and key. Billingsley expressly instructed David Jones, his miller at Nantgarw, in the preparative pulverising and grinding of the calcined bones to produce a very fine sieved bone ash that he considered suitable for incorporation into his paste mixture. He had experience of working in other china factories that used bone ash, such as Derby and Pinxton, and he would have seen the variability of the raw materials supplied to the works for inclusion in the porcelain pastes and the effect of using inferior quality materials. There was a hierarchical list of acceptable bones for calcination and ox bones came out top of the list, closely followed by cow bones, but horse bones were distinctly un-favoured: even so, their calcination had to be carried out in a prescribed way to avoid potential contamination from organic residues which could be deleterious to the porcelain during firing in the kiln—especially in the formation of carbon dioxide bubbles through decomposition of residual organic moieties in the setting paste during the kiln firing process, which would result in intercalated voids affecting the translucency of the incipient porcelain. To appreciate the importance of bone ash in the Nantgarw porcelain paste it is useful to look at the raw material compositions as given to Taylor (*The Complete Practical Potter*, Shelton 1847), who ascribed a recipe given to him by Samuel Walker, Billingsley's kiln manager, which comprised an initial mixture of 26 lbs ground bone ash, 14 lbs Lynn sand and 2 lbs potash mixed with water which was fired in a biscuit kiln. Then the 42 lbs of this cooled and finely ground frit were mixed with 20 lbs china clay to form the Nantgarw paste. Hence, the Nantgarw paste mixture comprised some 43% bone ash in its composition, clearly exceeding the percentage compositions of the other major components cited, namely sand and china clay. Bone ash when calcined is best represented as calcium hydroxyapatite, whose precise chemical formulation was disputed for many years until the seminal studies of Morgulis and Janacek (1931) definitively assigned its correct formulaic interpretation; the reader who wishes to learn more of these studies and the controversial background to the bone ash problem analytically is referred to the relevant chapters in a recent publication directed towards the analytical interpretation of Nantgarw and Swansea porcelain chemical data (Edwards, *Nantgarw and Swansea Porcelains*: *An Analytical Perspective* 2018). The calcined bone ash underwent further dehydroxylation during the high temperature firing process in the kiln to form the mineral whitlockite, a tricalcium phosphate, $Ca_3(PO_4)_2$, which is the species detected analytically in Nantgarw porcelain and from which back-calculation is necessary to determine the %age of bone ash in the original recipe. Suffice to conclude here that the incorporation of bone ash as a secret ingredient to Nantgarw porcelain was so well preserved by Billingsley and Walker whilst they were at Nantgarw that William Weston Young had absolutely no idea of its existence and importance when he undertook his own trial experiments to recreate porcelain in Nantgarw during the period 1819–1821. The presence of elemental phosphorus or molecular ionic phosphate detected analytically in porcelain paste is uniquely indicative of soft paste porcelain and bone china, as no other component raw material contains phosphorus. Hence, we would here find a definitive analytical indicator and discriminator between the porcelain made during the regime of William Billingsley and Samuel Walker and that of William Weston Young at the Nantgarw China Works as the former contained bone ash and the latter did not.

The chemical composition of the raw materials which comprise the Nantgarw paste and the minerals and compounds into which they are converted upon firing at the high temperatures required in the kiln are summarised in Table 9.1.

Table 9.1 Chemical composition of the raw materials comprising the Nantgarw Porcelain Formulation and the chemical species they are converted into upon firing in the kiln

Raw material	Chemical formulation	Kiln conversion	Final Porcelain material
Sand	SiO_2	SiO_2	Tridymite, Cristobalite
(Quartz, Silica)			
Flint	SiO_2	SiO_2	Tridymite, Cristobalite
Kaolinite	$Al_2Si_2O_5(OH)_4$	$CaSiO_3$	Wollastonite
		$CaNa[AlSi_2O_8]$	Bytownite
		Al_2SiO_5	Mullite
		$NaAlSi_3O_8$	Albite
Potash/Pearl ash	K_2CO_3		
Soda ash	Na_2O		
Lime	CaO		
Bone ash	$Ca_5(OH)(PO4)_3$	$Ca_3(PO_4)_2$	Whitlockite
Feldspar	$KAlSi_3O_8$	$Ca_2Al_2Si_2O_8$	Anorthite
(Soaprock)		$KAlSi_2O_6$	Leucite
		$KNaAlSi_3O_8$	Sanidine
Anatase	TiO_2	TiO_2	Rutile

References

J.W. Anthony, R.A. Bideaux. K.W. Bladh and M.C. Nichols, (eds.), *Handbook of Mineralogy, Volume VII: Silica and Silicates* (Mineralogical Society of America, Chantilly, Virginia, USA, 1995)

E.A. Barker, *Pottery and Porcelain in the United States* (New York, 1893) p. 178

C.F. Binns, *The Potters Craft: A Practical Guide for the Studio and Workshop* (D. Van Nostrand, New York, 1922)

W.F. Broderick, An English Porcelain Maker in West Troy, vol. 5, no. 2 23, p. 23, (The Hudson Valley Regional Review, 1988, September)

Sir A.H. Church, *English Porcelain: A Handbook to the China Made in England During the 18th Century as Ilustrated by Specimens Chiefly in the National Collection* (A South Kensington Museum Handbook, Chapman & Hall Ltd., London, 1885 and 1894)

E.S. Dana, *Textbook of Mineralogy: An Extended Treatise on Crystallography and Physical Mineralogy*, 4th edn. revised by W.E. Ford (Wiley, New York and Chapman & Hall Ltd., London, 1955)

L.W. Dillwyn, *Notes on the Experimental Production of Swansea Porcelain Bodies and Glazes.* Made by Lewis Weston Dillwyn with Samuel Walker at the Swansea China Works Between 1815 and 1817. Presented to the Library of the Victoria &Albert Museum, South Kensington, London by John Campbell in 1920. Reproduced in Eccles & Rackham, *Analysed Specimens of English Porcelain*, 1922, see reference below

H. Eccles, B. Rackham, *Analysed Specimens of English Porcelain in the Victoria and Albert Museum* (Victoria and Albert Museum, London, 1922)

H.G.M. Edwards, *Swansea and Nantgarw Porcelains: A Scientific Reappraisal* (Springer, Dordrecht, The Netherlands, 2017)

H.G.M. Edwards, *Nantgarw and Swansea Porcelains: An Analytical Perspective* (Springer, Dordrecht, The Netherlands, 2018)

J. Haslem, *The Old Derby China Factory* (George Bell, London, 1876)

M. Hillis, The development of Welsh porcelain bodies, Chapter 9. edn. J. Gray, *Welsh Ceramics in Context*, vol. II (The Royal Institution of South Wales, Swansea, 2005)

L. Jewitt, *Ceramic Art in Great Britain* (Virtue & Co., London, 1878)

S. Morgulis, E. Janacek, Studies on the chemical composition of bone ash. J. Biol. Chem. **93**, 455–466 (1931)

E.M. Nance, *The Pottery and Porcelain of Swansea and Nantgarw* (Batsford, London, 1942)

E.G. Ramsay, W.R.H, Ramsay, *Bow: Britain's Pioneering Porcelain Manufactory of the 18th Century.* (The International Ceramics Fair & Seminar, Park Lane Hotel, London, 2007, June 16) pp. 1–16

W.R.H. Ramsay, E.G. Ramsay, *The Evolution and Compositional Development of English Porcelains from the 16th C to Lund's Bristol c. 1750 and Worcester c. 1752—the Golden Chain* (Invercargill, New Zealand, 2017)

S. Shaw, *The Chemistry of the Several Natural and Artificial Heterogeneous Compounds Used in Manufacturing Porcelain Glass and Pottery* (Scott, Greenwood & Son, London, 1837). Re-issued in its original form in 1900, 713 pp, 1900

J Taylor, *The Complete Practical Potter* (Stoke-upon-Trent, Shelton, 1847)

J. Clegg, *The Pottery Gazette,* Vol. IX, no. 92 (Organ of the China & Glass Trades, Stationers' Hall, Ludgate Hill, London, 1885)

W. Turner, *The Ceramics of Swansea and Nantgarw* (Bemrose & Sons, Old Bailey, London, 1897)

Chapter 10
The Nantgarw China Works Site and Excavated Porcelain Shards

Abstract A history of industrial archaeology and the excavation of the Nantgarw China Works site, including the kilns and discovery of the waste pit and the acquisition of porcelain shards.

Keywords Nantgarw China Works site · Waste pit · Porcelain shards · Soft paste porcelain earthenware · Hard paste porcelain · William Billingsley · Samuel Walker · William Weston Young · Isaac Williams · Industrial archaeology

After closure of the porcelain manufactory at Nantgarw in 1823, the site remained derelict for a decade until William Henry Pardoe, who had assisted his father, Thomas Pardoe, in the tenure of the Nantgarw China Works with William Weston Young, re-opened the site in 1833 and started to manufacture pottery there, comprising utilitarian earthenware vessels and clay tobacco pipes, until his death in 1867. The business thrived and was adopted by his descendants until the 1920s, when eventually it fell into decline and was closed down. The Pardoe family continued to live for a while in the dwelling house, Tyla Gwyn, which had been the original residence of William Billingsley and Samuel Walker at the Nantgarw China Works Clearly, the site had rapidly fallen into decay over the preceding century and in February 1931, the National Museum of Wales undertook to investigate the site quantitatively and to excavate selected parts "*to ascertain by excavation particulars of the porcelain and earthenware constituting the genuine Nantgarw productions*". The man who was placed in charge of this excavation was Isaac Williams, Keeper of Art at the National Museum of Wales, Cardiff, and in 1932 he published his archaeological report in the learned journal, *Archaeologica Cambrensis*: this report was reprinted in a special publication of the National Museum of Wales in 1932, entitled "*The Nantgarw Pottery and its Products*: *An Examination of the Site*". To date, this report provides the first and most accurate picture of the original Nantgarw China Works, which was carried out within a decade of the final abandonment of the site.

It was immediately apparent that some perpetuated historical inaccuracies could be rectified for posterity: for example, Llewellyn Jewitt (*The Ceramic Art of Great Britain*, 1878) maintained that "*in 1823, the greater portion of the (Nantgarw) China works was pulled down, the dwelling house and some other portions alone remaining*". This conclusion is patently incorrect as Isaac Williams found that Kiln No. I,

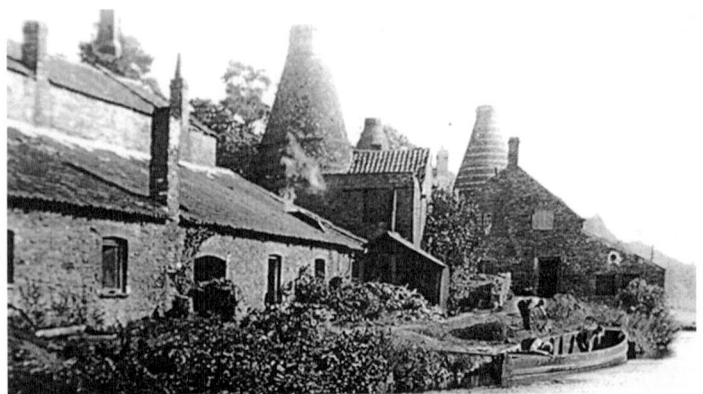

Fig. 10.1 A photograph of the Nantgarw China Works, taken around 1900 when it was still operational showing its prime location adjacent to the Glamorgan Canal: three kilns and several buildings can be seen, which expose the incorrect assumption of Llewellyn Jewitt in 1878 that the Nantgarw China Works was demolished after the auction sales in 1822–1823. Courtesy of the Nantgarw China Works Museum Trust, Tyla Gwyn, Nantgarw, Glamorgan

which he attributed to the biscuit porcelain firing kiln of Billingsley and Walker, and Kiln No. II, the glost kiln of Billingsley and Walker, were still standing—and, in addition, a third Kiln, No. III, which he claimed fired the earthenware and clay pipes of William Henry Pardoe in the mid-19th Century was also extant and still standing in 1932! Other buildings which constituted the original Nantgarw China Works and the later Nantgarw Pottery were also standing which were easily ascribed to specific ceramics processing functions, such as the potting room, drying room with racks and tables, a mill, a well-lit studio workshop, the pug-mill for density compression of the porcelain paste, a throwing shed and a store-house for the finished articles—all of these were still standing in 1932. It is obvious, therefore, that Jewitt had probably never seen the site but had believed earlier apocryphal statements that John Rose, proprietor of the Coalport China Works, had effectively dismantled the Nantgarw operation after his purchases at the auction sales and transported all the hardware, moulds, kilns and stock to Coalport in 1823. If this was indeed the case, there would literally have been nothing left for William Henry Pardoe to commence his start-up operation in 1833 at Nantgarw and certainly there would not have been evidence of the three ceramic kilns and associated ceramics working facilities which were still standing in 1932. Isaac Williams commented that the construction and integrity of Kilns I and II was apparently better than that of Kiln No. III, which accounted for their better state of preservation even allowing for their greater age in his opinion: in fact, Kiln No. III, which he attributed to William Henry Pardoe, was already starting to fall down in 1932! A very relevant photograph of the Nantgarw China Works site is shown in Fig. 10.1, which would have been taken around 1900 and it shows the factory in operation under the later Pardoe regime, with a small barge moored alongside the factory on the Glamorgan Canal and three kilns still standing amongst extensive outbuildings: hence, it can be concluded that Jewitt and other observers were obviously mistaken in the assumption that the Nantgarw auction sales had resulted in removal of all the hardware from the site and a dismantling of all of the operational buildings had thence occurred.

10.1   Industrial Archaeology of the Site

The science of industrial archaeology is a rather novel and quite recent branch of established archaeological science in that its origin and inception can be traced back in the UK effectively only to 1962, when the proposed demolition of the Euston Arch, Fig. 10.2, which was constructed in 1839 as a portal for the Great North Western Railway terminus in London, was carried out to make way for an extended railway passenger meeting hall and concourse despite intense public protest which had generated celebrity intervention and support, such as that of the Poet Laureate, John Betjeman. Nevertheless, the demolition of the Euston Arch went ahead as planned, but it was soon realised that in fact its demolition had been unnecessary and was avoidable for the planned British Rail terminus extension. The seeds were thus sown for the preservation of cultural heritage to be extended to industrial sites of national and international importance and since that time, the industrial heritage of abandoned factories and mines has been carefully evaluated and monitored for preservation and

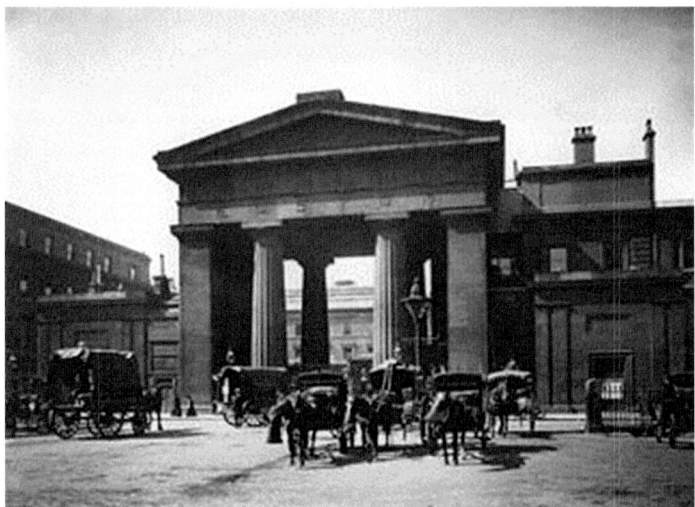

Fig. 10.2 The famous Euston Arch, terminus of the London and North Western Railway, erected in 1829, which was demolished in 1962/3 to facilitate the expansion of the concourse at Euston station for British Railways, amid much unsuccessful public objection and protest. Although the demolition proceeded, this action eventually gave birth to industrial archaeology projects and the awareness of the need for the preservation of industrial cultural heritage worldwide

archaeological excavation in a number of countries. However, there are some major problems associated with industrial archaeology, several of which can be identified with the Nantgarw China Works site under particular consideration here: the original purpose of an industrial site or factory can be obscured by later changes of operational use and the subsequent identification of materials and products pertaining to each phase of the original industrial operation can be made difficult thereby and obscured by later development. This is particularly relevant to the discovery of broken items in waste tips or dumps, which may have no relevance to the archaeological context or temporal period of archaeological interest. Likewise, the importation or acquisition of special materials and items belonging to later periods of factory operation can be found mixed with earlier possible evidential material: in the case of an abandoned site, its usage as a depository or dump for waste materials from another, perhaps completely unrelated site, is also appreciated and these may have no relevance or archaeological context at all for the site under excavation or study.

At Nantgarw, it is very fortunate that the forethought and rapid action of the National Museum of Wales in 1931 enabled the first measurements and archaeological excavation of the China Works site to be undertaken by Isaac Williams within a decade of its abandonment in the 1920s and before irrelevant material from other sites had been accumulated or deposited there. The discovery of the waste dump and removal of specimens of porcelain and earthenware shards is first recounted in Isaac Williams' report (*The Nantgarw Pottery and its Products: An Examination of the Site* 1932), in which he provides useful and precise details of the location of his sampling trenches and a cross-section of the waste tip trench which was sunk through a pronounced depression situated 30 feet east from the eastern wall of the potting shed: the five archaeological strata he was able to identify in this trench comprise sequentially a vertical transect of some 4″ of turf and topsoil, 6″ containing fragments of pipes, stoneware and earthenware, 10″ of earthenware, stoneware, *Rockingham*-type teapots and pipes, a further 6″ of stoneware, some porcelain fragments and pieces of saggars, and finally 12″ at the lowest level containing porcelain shards and a few saggars. Beneath this lowest level there was virgin soil. A diagram of this waste tip vertical transect constructed by Isaac Williams and reproduced from his archaeological report is shown in Fig. 10.3.

Hence, we can conclude that the porcelain shards from the William Billingsley, Samuel Walker, William Weston Young and Thomas Pardoe combined operations at the Nantgarw China Works site will be located in the lowest level of this waste tip, that is at level 5, some 38″ (approximately 1 m) below the surface. Above this, that is in levels 4-1 vertically in the transect, the fragments comprise assorted earthenware and clay pipe wasters only with the exception of a few porcelain pieces found in the fourth level. The discovery of porcelain fragments in the level four above the Nantgarw China Works Phase I and II operations and mixed in with earthenware fragments could possibly be significant as it implies that porcelain was manufactured at Nantgarw along with earthenware items by William Henry Pardoe? It seems that Isaac Williams did not identify separately the location of each shard excavated—this information would have been invaluable in a forensic archaeological context for the discrimination between later productions and their analogues from the earliest phase

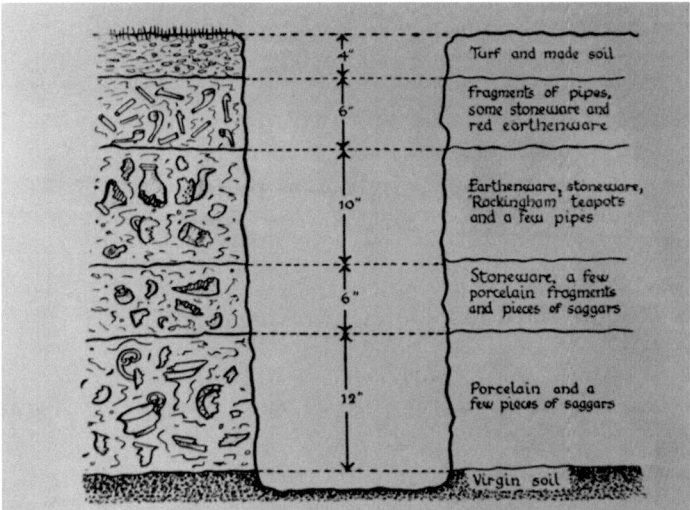

Turf and made soil

fragments of pipes, some stoneware and red earthenware

Earthenware, stoneware, 'Rockingham' teapots and a few pipes

Stoneware, a few porcelain fragments and pieces of saggars

Porcelain and a few pieces of saggars

Virgin soil

Fig. 10.3 The Nantgarw China Works waste pit discovered by Isaac Williams during his 1932 excavation carried out at the site: this waste tip lay some 30 feet east of the eastern wall of the potting shed. Five distinct levels were identified by the archaeologists in this waste pit—the porcelain shards from the Billingsley/Walker/Young period occupied Level 5 at a depth of approximately 38 inches (~1 m) below the top surface, dating from 1817–1820, which formed the basis of the modern analytical studies undertaken here and described in this work. The discovery of buildings and kilns ascribed to Billingsley and Walker was at first surprising because William Chaffers and Llewelyn Jewitt in the mid- to late-19th Century had maintained that the site was destroyed after Young left Nantgarw in 1823 and all useful equipment had been removed to the Coalport China Works

of operations—but an intermingling of waste porcelain shards with earthenware, which was never made at Nantgarw in the Billingsley/Walker/Young period does give rise to this intriguing idea of a later porcelain production and therefore lends it credibility. We have seen already that William Henry Pardoe did announce the re-opening of the Nantgarw site for porcelain production in 1858, but that he also ceased this activity soon afterwards and that he advertised its disposal through an auction sale. It is of course difficult to estimate if his commercial production of porcelain ever succeeded or perhaps was discontinued soon after his trials.

Isaac Williams also remarked that of the three standing kilns at Nantgarw, Kiln I, which he assumed was the porcelain biscuit main firing kiln, Kiln II was the glost kiln from the Billingsley and Walker era and a rather decrepit Kiln III would have been the earthenware firing kiln from the Pardoe Jr. era, and that this was rather poorly constructed in comparison with the other two. In this context, a later excavation by the Gwent and Glamorgan Archaeological Trust in the 1990s was directed at the dissembling and re-erection of Kiln II, during which the remains of clay pipes and earthenware shards were discovered built into the kiln structure, demonstrating that this, in fact, should now be better assigned to the earthenware Pardoe kiln and not the Billingsley/Walker glost kiln as had been assumed by Williams. Unfortunately, the details of this more recent excavation and the detailed archaeological report associated with it have now been lost, having been submitted to the Rhondda—Cynon Taff County Council following the site study, and no trace of this report can now be located (Nantgarw China Works Museum Trust, private communication). However, a scientific paper was presented afterwards to a learned journal (Murphy et al. *Post Mediaeval Archaeology* 1997) on the excavation which provides the source material for their conclusions and later interpretation.

Isaac Williams (*The Nantgarw Pottery and its Products: An Examination of the Site,* pp. 19–20 1932) also states that he found it significant that only one or two small fragments of glazed porcelain were found in his archaeological excavation of the shards and none at all showed signs of any enamelling decoration. The precise location of the excavation of the shards within level 5 of the waste pit is now lost, but from their composition, type and location they clearly belong to the Billingsley and Walker era and can be confidently be assigned to the lowest level 5 of the Williams' vertical transect stratification and because of the absence of any associated earthenware relict material. Many of the shards discovered in level 5 of the waste pit have shapes which match finished and marked pieces of Nantgarw porcelain of the Billingsley and Walker period, *ca.* 1817–1819.

From his archaeological excavations, shard examination and site inspection Isaac Williams concluded that:

- there were no losses of biscuit porcelain adjudged fit to be glazed;
- great care must have been exercised in the preservation of decorated examples and that the number of damaged, decorated specimens must have been very few in Billingsley and Walker's time;
- the export to London of the major portion of the Nantgarw glazed ware in the white during the years 1817–1819 is substantiated;
- this ware consisted chiefly and mainly of dinner, dessert and tea services

William Weston Young in his production trials of porcelain carried out between 1819 and 1822 appears to have used a composition of paste similar to that invented by Billingsley;

Young used the Billingsley moulds for pressing and copied the Billingsley models for throwing, but when he did attempt anything new as a potter it was generally thicker and heavier in construction.

Several remarks can be made about the veracity of these conclusions in the light of modern research: firstly, it is a reasonable assumption that almost all of the Nantgarw porcelain that was saleable and fit for purpose from the Billingsley and Walker era would have been sent to London for decoration—with a 90% kiln wastage in the biscuit stage, every piece of Nantgarw porcelain that was perfect would have been despatched by Billingsley to John Mortlock's outlet for sale in the capital, as demanded by Mortlock. Small defects in some pieces, it has been alleged, would have destined them to be kept for decoration for the local market, but even then, we should perhaps query whether a discerning master such as Billingsley have been content to decorate porcelain personally that was in any way inferior in quality? It is true that other perhaps younger and less accomplished artists would have been engaged upon this exercise and Richard Millward, who was employed as a young man at the Nantgarw China Works, has enumerated several of these in his statement to William Turner (*Ceramics of Swansea and Nantgarw* 1897) about the Nantgarw workforce, for example, May Hewitt and possibly Lavinia Billingsley. Secondly, it has been noted elsewhere that occasionally Nantgarw pieces are noted with a seemingly poorer quality glaze—and that shards have been found which are seen to have been glazed but that the glaze is pitted, which has been attributed to excessive gas evolution during firing of the piece in the glost kiln—this indeed would have been a grave loss as the piece concerned would have been judged previously to have been one of the surviving 10% of pieces in biscuit porcelain that was viable for sale. The fact that so few decorated and enamelled shards are found (*sic.*, "*none*" according to Williams) indicate that having passed through the biscuit firing stage then few losses would have been encountered during the subsequent glazing and enamelling operations for despatch to London or for local consumption. The basic output of the factory would have been a "service" for tea, dessert and dinner and Nantgarw ornamental or specialised items are only rather rarely found: examples include spill vases, *pot pourri* jars, writing sets, cabinet cups, taper sticks and inkwells. These often as a result now command very high prices for their rarity at auction. Perhaps the most contentious issue relating to Isaac Williams' conclusions are the last two points he makes, which assume that William Weston Young actually made porcelain at Nantgarw using William Billingsley's paste recipe and formulation. It is clear that Isaac Williams did not have access to Young's work notes of his experiments in trying to recreate Nantgarw porcelain, from which it is obvious that William Weston Young never had any detailed knowledge of Billingsley's paste and he certainly did not realise the importance of the bone ash component, which he consequently omitted entirely from his formulation recipes. Hence, Young, if he did manufacture porcelain at Nantgarw between 1820 and 1822 (note the revised dates, as Billingsley and Walker were still working and producing porcelain at Nantgarw in 1819!!) could not have created a Billingsley/Walker body. Tantalisingly, Williams does not back up his statement in this regard with the discovery of shards that he can undisputedly attribute to William Weston Young's experiments—it must be concluded that, if in fact Young had managed to recreate a "Nantgarw" porcelain in 1820–1822 even approaching the original formulation then he would also have been faced with the appallingly high kiln losses as had Billingsley and Walker, so there would be a significant number of shards existing at the top of level five or the bottom of level four in Williams' excavation of the waste dump to reflect this. However, if Young had experimented with a different type of Nantgarw porcelain paste, which may have been successful in his trials if not commercially, there should be shards to indicate this somewhere in the waste pit and we would expect to locate these at the juncture of levels four and five and at a transect depth greater than that where the earthenware fragments are found as they would naturally have pre-dated the efforts made by William Henry Pardoe to re-open the factory for the manufacture of earthenware in 1833. It must be remembered also that Young did not have the services of a competent kiln engineer of the merit of Samuel Walker at his disposal.

10.2 Nantgarw Porcelain Shards: Their Origin and Analytical Presentation

The first published chemical analyses of Nantgarw porcelain reported were accomplished successfully using damaged, finished and decorated pieces by Eccles and Rackham in (1922) from two specimens, namely a saucer and a soup dish, from the Lady Charlotte Schreiber Collection housed in the Victoria and Albert Museum, South Kensington, London (Eccles and Rackham, *Analysed Specimens of English Porcelain* 1922) and a further specimen from a marked and broken Nantgarw plate which was also housed there at that time but which did not appear in their analytical resume (this was seen and recorded by Maurice Hillis and the analytical data reproduced in his article in *Welsh Ceramics in Context*, Chap. 9, 2005): these analyses were accomplished using a wet chemical acid digestion of relatively large samples of Nantgarw plates and saucers and were, therefore, totally destructive of the samples. The analyses necessarily included elemental oxides found in the surface glaze as it was not possible to undertake a differential sampling between the glaze and the body. One of these samples was a tea cup and saucer from the renowned *Duncombe* service which had been produced at Nantgarw and decorated locally by William Billingsley with garden roses: it was presented to Edward Edmunds, landlord of the Nantgarw China Works site, and had been given to him personally by William Billingsley in *lieu* of rent. It was therefore part of a highly desirable and historically documented service, finding its way eventually through the marriage of one of Edmunds' daughters into the estate of the Reverend Duncombe of Llandaff Cathedral, whence it acquired its name (Edwards, *Swansea and Nantgarw Porcelain: A Scientific Reappraisal* 2017). Sir Arthur Church claims to have performed analyses of Nantgarw porcelain some years earlier in his book *English Porcelain,* which was published in (1894) following his Cantor Lectures (Church 1881), but, unfortunately, he does not provide details of the quantitative component elemental oxides found therein such as silica, alumina, lime, magnesia and phosphorus pentoxide, unlike the porcelains of other factories he quoted there, such as Chelsea and Bow, or of the Nantgarw samples he used although he does illustrate a saucer in the Lady Charlotte Schreiber Collection at the V&A (South Kensington) Museum (Fig. 48, facing p. 96) and refers to twelve other specimens in the Jermyn Street Collection, originally acquired by the distinguished geologist Sir Henry de la Beche (de la Beche 1876). All three of the Eccles and Rackham specimens were fragments of broken porcelain, all of which were glazed, two of which were decorated with coloured enamels; two of these specimens are still housed in the Victoria and Albert Museum, London, archive and the third now resides in the Nantgarw China Works Museum, where it was lodged via the National Museum of Wales, Cardiff. Although only the *Duncombe* service saucer is illustrated in Eccles and Rackham's monograph (*Analysed Specimens of English Porcelain* 1922, coded C586-1919, No. 22 on Plate VIII) all three are illustrated in Hillis' article (*Welsh Ceramics in Context* 2005): the second of these, analysed and reported by Eccles and Rackham, fully described there as C510-1919, No. 21, as a broken soup dish, undecorated and glazed in the white which is not illustrated, and the third is shown as a badly damaged plate with enamelled floral decoration for which the analytical data were recorded in Eccles' manuscript notes now in the possession of Jonathan Gray (Ref. Hillis, *Welsh Ceramics in Context*, volume II 2005). All three of these are illustrated in Hillis' article in plates 9.12, 9.13 and 9.13 on p. 177 and 178, and the analytical data are presented for them as specimens labelled E1, E2 and E3, being the *Duncombe* saucer, the white soup plate and the enamelled decorated plate, respectively.

Some 70 years later, Tite and Bimson (1991) reported the analytical data from three specimens of Nantgarw porcelain shards which were stored in the British Museum archive. As there is no record of any subsequent archaeological excavation being undertaken after Isaac Williams and that of the National Museum of Wales in 1931 which could have deposited shards recovered in the British Museum, then it could safely be assumed that these shards actually came from the Williams excavation and had then been placed in the British Museum archive. However, in an acknowledgement at the conclusion of their paper, Tite and Bimson record their gratitude to Dr Mark Pollard, then of the University College of Wales, Cardiff, for the gift of the three Nantgarw shards used in their analyses, despite their statement in an opening paragraph of their paper that all shards were taken from the British Museum Research Laboratory archive.

The Tite and Bimson paper was followed in 1998 by a study of Owen et al. of ten more Nantgarw shards and a year later by Owen and Morrison, in 1999, of a further 9 Nantgarw shards and one finished and decorated piece of Nantgarw porcelain, although badly damaged, numbered N37: the former ten specimens were found at the site of the Nantgarw China Works Museum, whereas the second batch of nine shards were provided from the Nantgarw China Works Museum site, a Nantgarw plate from the China Works Museum providing the decorated specimen, although it is seems that this badly damaged plate (Hillis, *Welsh Ceramics in Context,* Volume II 2005, colour plate 9.12, p. 177) is identical with that analysed by Herbert Eccles and Bernard Rackham in (1922), coded E3 by Hillis, and described earlier. Owen and Morrison also state in their second paper that the ten Nantgarw shards were found "*lying around the site*"—which does not really confer upon them any significant archaeological depositional or burial context either for their location or correct chronological placement. A full listing of the Nantgarw shards and their origins which have been analysed thus far in the literature is given in Table 10.1 and later full analytical details are provided in subsequent tables, for the bodies, glazes and the pigments on decorated shards. The claim by Sir Arthur Church to have analysed two complete specimens of Nantgarw porcelain in 1894 is recorded in the Table but no further assessment can be made in the absence of the analytical data: it is known that Church did give some lectures around this time (the Cantor Lectures, see references) and it is possible that the analytical data were presented there verbally to the audience but the author can find no record to this effect.

Hence, in total, only 33 shard specimens of Nantgarw porcelain have ever been studied and reported thus far in the literature, in addition to four finished and decorated, but damaged specimens acquired from Museum collections. In the 1990s, it is recorded that the Rhondda-Cynon-Taff and the Gwent and Glamorgan Archae-

Table 10.1 Origin of Nantgarw Shards and Porcelain Specimens Subjected to Analytical Study in the Open Literature

Investigator	Type and origin of specimen	No. of specimens	Conclusion
Church (1894)	Finished porcelain/V&A Museum*	2	?
Eccles and Rackham (1922)	Finished porcelain/V&A/NCW Museum	3	Soft paste
Tite and Bimson (1991)	Shards/BMRL**	3	Soft paste
Owen et al. (1998)	Shards/NCW site	10	Soft paste + hard paste
Owen and Morrison (1999)	Shards + Finished porcelain/NCW Museum***	9/1	Soft paste + hard paste
Edwards and Colomban (2019)	Shards/NCW Trust****	11	Soft paste + hard paste

*Victoria and Albert Museum, South Kensington, London
**British Museum Research Laboratory, Bloomsbury, London
***Nantgarw China Works Museum, Nantgarw
****Nantgarw China Works Trust, Nantgarw

ological Trust set about the restoration of the biscuit porcelain kiln assumed used by Billingsley and Walker between 1817 and 1820 (Fig. 10.4). During this investigation, a large number of shards were again recovered from the waste pit and very recently, in 2018, a scientific analysis of some of these shards has been undertaken by the author in collaboration with Professor Philippe Colomban at the Laser Micromole Laboratory in the University of Paris, using SEM/EDAXS elemental determinations and Raman molecular spectroscopy for the first time. These particular Nantgarw shard analyses will be identified in the Table by the code "E&C 2018".

All the earlier data comprised an elemental analysis of the Nantgarw shard bodies with several later additions specifically for the glaze—whereas the earlier studies were undertaken using bulk wet chemical digestion procedures (Eccles and Rackham, *Analysed Specimens of English Porcelain* 1922), the later studies were made using scanning electron microscopy and electron dispersive X-ray spectroscopy (SEM/EDAXS) (Tite and Bimson 1991; Owen et al. 1998; Owen and Morrison 1999) on micro-specimens. In all these cases, the compositions are recorded in terms of their metal or non-metal oxides, such as silica, alumina, magnesia, lime, potash and soda. A further specimen found at the site by Owen and Morrison was a fragment of glass, but the analysts comment quite correctly that this could perhaps be "associated" and was not necessarily a specimen of any raw material used in the Nantgarw porcelain recipe formulation (it is already known that Billingsley did not use glass frit in his body paste recipe from the detailed list of components provided

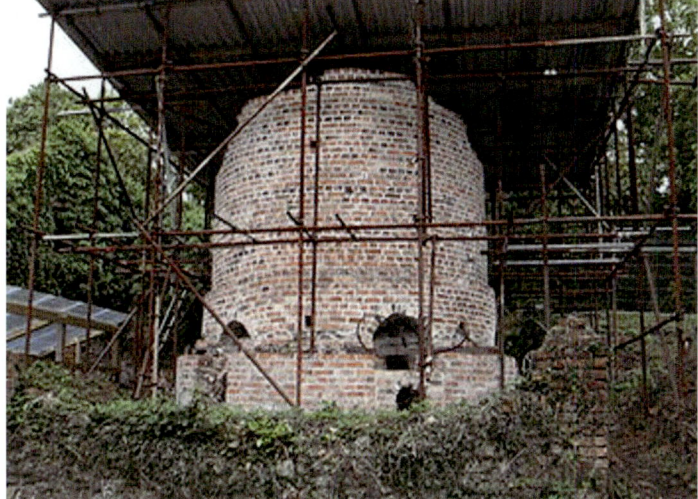

Fig. 10.4 A restored biscuit kiln from the Billingsley and Walker period undertaken by the Rhondda-Cynon-Taff and the Gwent and Glamorgan Archaeological Trust in the 1990s. Courtesy of the Nantgarw China Works Museum Trust, Tyla Gwyn, Nantgarw, Glamorgan

by Samuel Walker to John Taylor and published in 1847 in *The Complete Practical Potter*).

The recent combined collaborative analytical investigation referred to above has been carried out by Edwards and Colomban (2019): these analyses were carried out using 11 shards from the Nantgarw China Works Museum archive, of which three bore traces of enamelling in various colours on glazed shards, using SEM/EDAXS spectrometry, X-ray diffractometry and laser Raman spectroscopy—all of which provide elemental and molecular information about the porcelain bodies and, where appropriate, the glazes and pigments that have been applied to them. The shards comprised moulded and unmoulded fragments of plates and dishes, glazed and unglazed specimens, sagged porcelain and three shards which showed evidence of enamelling and glazing—a type of porcelain reject which Isaac Williams claims not to have seen anywhere in the waste tip site—signifying that the major source of spoiled Nantgarw porcelain occurred at the initial firing and post-firing primary glazing stages before enamelling was employed and prior to the subsequent glazing of the perfect pieces in the glost kiln. This increased the number of Nantgarw shards that have been subjected for analysis and which have been reported in the literature.

A summary of the origin of all Nantgarw porcelain shards studied thus far is given in Table 10.1 along with the authors of the scientific papers describing their analyses and conclusions.

10.3   A Prophetic Prediction

Ernest Morton Nance (*The Pottery and Porcelain of Swansea and Nantgarw*, p. 390 1942) has commented that:

> Nantgarw porcelain can then potentially be made today of the same fine quality as originally made by Billingsley. But whether any manufacturer is likely to adopt the process is another matter; for, unfortunately, the same difficulty which caused the failing of the Nantgarw works still remains to be overcome. Mr. Bott (Author: T.J. Bott, Art Director of the Coalport China Works) has informed me that at Coalport no porcelain body has to his certain knowledge been made from the Nantgarw recipe even on a trial since about 1860, and that the clay is deficient in the plastic ingredients which go to make a workable china body and is not likely to be made there. Perhaps, in the course of time we shall once more see this beautiful porcelain reproduced, at all within small quantities, as an article of luxury for the gratification of those who appreciate a beautiful and distinctive porcelain. But it will probably prove to be too expensive ever to become an ordinary article of trade.

Sequel: Nantgarw porcelain has now been successfully reproduced commercially for the first time in 2018 after two centuries by a consortium, with financial support through a development grant from the Welsh Office, using Billingsley and Walker's body recipe and a derived glaze formulation in 2018, to celebrate the 200th anniversary of the first production of Nantgarw porcelain by Billingsley, Young and Walker. Observers agree that this recreated porcelain is an almost exact match for the original Nantgarw porcelain and its successful manufacture owes much to the advances made in our understanding of the high-temperature chemical interactions which occur between the components, the sourcing of analogous raw materials, their careful preparation and mixing and modern improvements in kiln firing control technology and stability. Examples of two modern "Nantgarw porcelain" cabinet cups are shown in Fig. 10.5, which illustrates well the superb translucency of this fine soft paste porcelain china: as part of the current 2018 analytical investigation, several unglazed biscuit shards of this. Nantgarw 200th anniversary china, code-named NArt200, were also made available for comparative study by Edwards and Colomban (2019) and the results are compared here against the original Nantgarw porcelains of Billingsley, Walker and Young from some two centuries earlier.

10.4   Nantgarw Porcelain Shard Analyses

11 selected shards of Nantgarw porcelain were analysed at the LASIR laboratories of the Universite de Pierre et Marie Curie, Paris, by Professor Philippe Colomban et al., using laser Raman spectroscopy (LRS), scanning electron microscopy, X-ray Diffractometry (XRD) and electron dispersive X-ray spectrometry (SEM/EDAXS) in collaboration with the author. These shards included enamelled and glazed fragments. The occurrence of the latter in the Nantgarw waste pit must be considered rather rare in comparison with other shards, glazed or unglazed, as they represent the unsatisfactory production of porcelain which had survived the initial firing in the

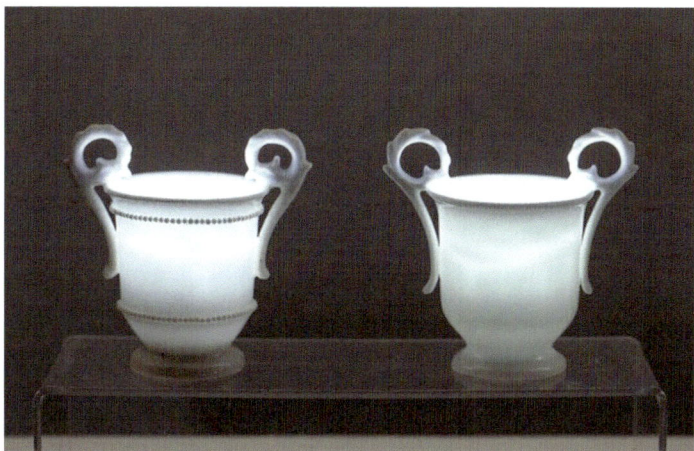

Fig. 10.5 Two cabinet cups of NArt200 porcelain (i.e. Nantgarw 200th anniversary), made in 2018 to celebrate the bicentenary of the opening of the Nantgarw China Works for commercial production in 1818, using the Nantgarw recipe quoted in John Taylor's book (*The Complete Practical Potter* 1847) allegedly provided to him by Samuel Walker before he departed for the USA to set up a pottery at Temperance Hill, New York State. Courtesy of the Nantgarw China Works Museum Trust, Tyla Gwyn, Nantgarw

biscuit kiln and would have been damaged in the final stage of glazing at lower temperatures in a glost kiln, after enamelling, when firing losses would have been anticipated to be very few in number.

A summary of the scientific data and their interpretation from the combined Raman spectroscopic molecular and elemental analyses of the author and Professor Philippe Colomban now follows.

Of the range of Nantgarw excavated shards examined by Raman spectroscopy and SEM/EDAXS spectrometry in the UPMC Paris LASIR laboratories, one shard in particular excites especial interest because of the different spectral signatures it exhibited. Normally, a Nantgarw porcelain body will show the Raman spectral (LRS) signatures of its high-temperature quartz polymorphs, cristobalite and tridymite, along with phosphatic signals from the apatite component, whitlockite, and silicaceous materials such as wollastonite and leucite. However, the characteristic phosphatic signatures expected do not appear at all for shard number NG6—a large, glazed fragment in the group. The conclusion, therefore, is that the shard NG6 does not contain any whitlockite or apatite and that it must therefore be devoid of bone ash in its component raw materials. Effectively, this means that NG6 comprises a hard paste porcelain body, unlike all the other specimens of Nantgarw shard porcelain examples tested here. SEM/EDAXS data confirm the LRS signatures and show that an elemental phosphorus signal is absent. Since this shard was excavated along with genuine Nantgarw porcelain shards from Level 5 in the waste pit, the inevitable inference can be made that it was made at the site but not by Billingsley and Walker! Hence, we can conclude that this fragment is quite probably a residue of the efforts of

William Weston Young to re-create his version of Nantgarw porcelain in 1819–1820 and thereafter, for which we have already seen that he was unaware of the addition of the vital bone ash component in Billingsley and Walker's recipe, as revealed by John Taylor in 1847 (*The Complete Practical Potter* 1847), the year that William Weston Young died. A footnote in Morton Nance (*The Pottery and Porcelain of Swansea and Nantgarw* 1942) comments on a strange piece of porcelain that had surfaced in the possession of Young's descendants in Bristol (actually, to be more correct, the relatives of John Wright, who supported William Weston Young in his literary endeavours at Bristol, culminating in the publication of his book *The Guide to the Beauties &c. of Glyn-Neath* 1835). In this account, Morton Nance inspected a "Nantgarw" plate that the family maintained was given to John Wright personally by William Weston Young and which had been decorated by him. Morton Nance concluded that the plate, although of a reasonably good texture and of an exceptionally clear translucency, was nevertheless not the quality of genuine Nantgarw porcelain, and that it was more likely to be a hard paste porcelain. Morton Nance, therefore, maintained that it was probably sourced from somewhere in Staffordshire. However, in the light of the discovery of the unusual shard NG6 during the present analyses, another possibility is now presented—that the plate possessed by John Wright's descendants in fact could be a specimen of Young's hard paste porcelain made in his trials at Nantgarw during 1819 and 1820? Confirmatory evidence of the quality of Young's porcelain body is given by the observation of the translucency of shard NG6, as seen in Fig. 10.6, and indeed it is seen to be excellent!

A reasonable conclusion that can be inferred from the preceding discussion, therefore, is that Young did succeed in his trial experiments to produce a hard paste porcelain body at Nantgarw, but to call this a re-creation of the original Nantgarw porcelain, although the product was actually made at Nantgarw, is stretching credibility too far. However, the importance of the shard NG6 is that it is potentially the first documented piece of evidence that Young did successfully fire his recipe at the Nantgarw China Works during his time there in 1819–1823 and as such it now becomes a very important historical ceramic specimen. An interesting feature of NG6 which may have escaped initial attention is that it has been glazed but not decorated: this implies that Young was pleased with his biscuit firing of the specimen which gave rise to the shard, but that something went sadly amiss in the glazing or glost kiln during its subsequent firing there. Perhaps there was evidence of ineffective glazing over the piece or perhaps a "glaze creep" had occurred in which the surface tension between the glaze slip and underlying body had caused a withdrawal of glaze from parts of the surface, resulting in an unsightly pooling of glaze over some regions of the specimen and a deficiency in others. Whatever the real reason, a very nice piece of porcelain with an apparently quite acceptable translucency had to be destroyed—contributing to the high level of kiln wastage that has continually beset the commercial enterprise at the Nantgarw China Works from the outcome. As will be discussed further later, another key analytical result from the shard NG6 is that it contains lead oxide in the glaze: therefore, it cannot be the same glaze used by Young and Pardoe in firing the residual biscuit porcelain remaining at Nantgarw from Billingsley and Walker, as Young's No. 2 glaze was lead-free! However, Lewis Dillwyn (Dillwyn Notebooks,

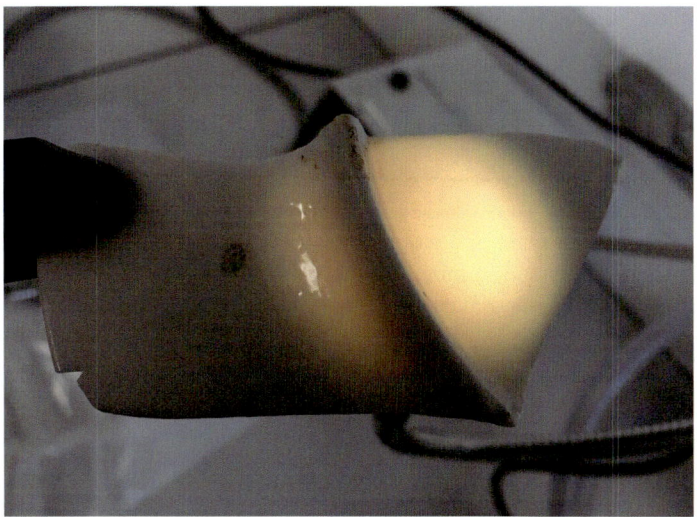

Fig. 10.6 Shard NG6 from the Nantgarw China Works set of shard specimens excavated from Level 5 of the waste pit. Photographed in transmitted light to demonstrate its excellent translucency, closely approaching that of the true Nantgarw soft paste body. This shard, from a dinner plate or soup dish, has already been glazed, so it must have passed inspection after firing in the biscuit kiln firing process and has thereafter suffered some fault in the glost or glazing kiln. Courtesy of Professor Philippe Colomban, LASIR laboratory, Universite de Pierre et Marie Curie, Paris

Appendix B; Eccles and Rackham 1922) had already noted, during his trial experiments in 1815–1817 that to perfect a more robust Swansea porcelain using a soaprock component at the expense of bone ash, the glaze formerly used for his duck-egg bone china was not compatible with the harder china and that an increased feldspar component was required along with flint glass—it seems that Young would have realised this also and had to perfect a new glaze for his prototype china to replace the No. 2 glaze that worked so well with the Billingsley and Walker body formulation.

10.5   Conclusions of Analytical Studies of Nantgarw Shards

The combination of molecular spectroscopy and elemental determination has revealed a potential method for the non-invasive interrogation of Nantgarw porcelain to facilitate the attribution of unmarked and specimens of uncertain origin: this can be accomplished readily and without the necessity of taking samples or in any way altering chemically or mechanically the integrity of the specimen, so it is applicable for the interrogation of perfect pieces of rare porcelain from this factory. In the context of this book, the discovery of shard NG6 which is a hard paste porcelain formulation along with the more characteristic soft paste porcelain shards that are assignable to the Billingsley and Walker recipe from Level 5 of the factory waste pit on the Nantgarw China Works site is conclusive evidence that William Weston Young did manage to create his version of "Nantgarw" porcelain there during his tenure there between 1819 and 1822/3. It is appropriate, therefore, to recount the facts surrounding this episode so that much of the historical record that has been promoted hitherto can be re-evaluated correctly.

Ernest Morton Nance, whose excellent book on the *Pottery and Porcelain of Swansea and Nantgarw* (Batsford 1942) encompassed some forty years' research on Welsh pottery and porcelain and whose meticulous digestion of the relevant documents and records has done so much to establish the facts surrounding the complex history and personalities associated with the Swansea and Nantgarw enterprises is occasionally found to be quite ubiquitous in several areas—and the manufacture of porcelain at Nantgarw by William Weston Young is a case in point. Some of the key features of the argument as proposed eloquently by Morton Nance, very often raised in subsidiary footnotes to other main textual themes, can be summarised below:

William Weston Young's Diaries, which Morton Nance has clearly perused and studied in great detail, and which he has elaborated from Young's shorthand notes, has occupied an Appendix to his book (*The Pottery and Porcelain of Swansea and Nantgarw* 1942) amounting to some 45 pages of closely spaced written entries, make no mention of his purchase of materials for porcelain manufacture. An exception, however, is provided by single, rather terse entries for the purchase of gold, necessitating his travel to Bristol, for gilding glazed porcelain and the acquisition of borax for an alkaline fusion component in the glaze. It is well known that Young experimented with the production of a glaze with which he could finish off the biscuit porcelain remaining at Nantgarw after the departure of Billingsley and Walker early in 1820. This so-called, *Nantgarw* No. 2 *glaze*, which is used for the re-firing and decoration of some of this residual porcelain between 1820 and 1823 is easily identified and is often referred to as a creamy *Pardoe glaze,* thinner and perhaps more pitted than that of the stark, white glaze used by Billingsley and Walker. Yet, in the same diary records, Young has itemised a bill payment made to David Jones, a local miller, for services rendered in the grinding and milling of porcelain components! So, clearly, William Weston Young was engaged in porcelain manufacture at Nantgarw, as we already know from his diary entries in 1819 and 1820 recounting his four potential formulaic mixtures being trialed for porcelain manufacture; it is interesting that Morton Nance (p. 407) dismisses these experiments as "*unsuccessful and failures*", seemingly without citing any evidence for this conclusion. He then closes this remark with an assertion that Young would have been engaged purely in decorating china and he would not really have had much time for that, given his other activities, yet Pardoe really bore the brunt of the decorative duties at Nantgarw between 1821 and 1823.

In an earlier footnote, Morton Nance (*The Pottery and Porcelain of Swansea and Nantgarw* 1942, p. 402) is intrigued to report on a communication from Messrs Steele and Garnett, whose home abutted onto the Nantgarw China Works site in the early 20th Century, had discovered fragments of a "*glassy porcelain*" in their garden, which they assumed would once have been a part of the China Works operation. He attributed these fragments, which were totally unlike the other Nantgarw shards

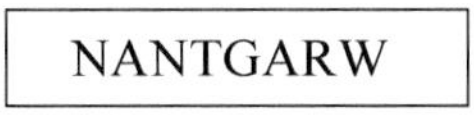

NANTGARW

Fig. 10.7 Diagram of the impressed Nantgarw mark apparently adopted for use by William Weston Young on his "hard paste" Nantgarw porcelain variety, contained within a rectangular box-like cartouche and missing the hyphen between the NANT and GARW. It is claimed that the impressed letters are smaller than those used by Billingsley and Walker for their Phase II Nantgarw China Works wares

usually found at the site, to experiments carried out by Billingsley and Walker, or perhaps even William Henry Pardoe later in the 1850s. Could these perhaps have originated from William Weston Young's trials?

A very curious *non-sequitur* appears in another Morton Nance footnote referring to the ease of identification of any porcelain that had been manufactured by William Weston Young at Nantgarw because it would have been painted by Thomas Pardoe, and since Pardoe's work is found only on Billingsley's Nantgarw paste body then Young could not have made any porcelain there! Morton Nance then asserts this must be the case because Herbert Eccles, after his inspection of the Nantgarw china on view at the Glynn Vivian Loan Exhibition at Swansea in 1914 to mark the centenary of the founding of the Swansea and Nantgarw manufactories, pronounced that in his opinion there was only one unique identifiable Nantgarw body—and that it was truly "*invariant*". The corollary to this statement was equally remarkable because Eccles then went on to say that if Young had succeeded in making porcelain at Nantgarw it would hence have been indistinguishable from that made by Billingsley and Walker! Yet, some twenty years earlier, Turner (*The Ceramics of Swansea and Nantgarw* 1897) reported that Young's porcelain was of an excellent quality, harder, but was stamped NANTGARW in a smaller impressed script, minus the hyphen, and which was contained in a boxed cartouche (A diagrammatic representation of this mark is shown in Fig. 10.7). Unfortunately, this statement has been ignored and cannot be verified by any cross-checking as to where Turner had seen and handled this mysterious and novel type of "Nantgarw" china.

Chronologically, a very curious situation exists surrounding Eccles' remarks made in 1914 as just a few years later in 1922, in his seminal work with Bernard Rackham (Eccles and Rackham, *Analysed Specimens of English Porcelain* 1922), the recipes and work book notes recounting Lewis Weston Dillwyn's experiments with Swansea porcelain, aided by Samuel Walker between August 1815 and late 1817, were reproduced in their entirety there (Appendix B). Yet, Eccles maintained that Young was trying to emulate Dillwyn's Swansea soaprock *trident* porcelain at Nantgarw and failed to do so; Morton Nance found this a strange development as he pointed out that clearly Eccles had not appreciated Dillwyn's extensive and authoritative notes, otherwise he would see that Young could not have been trying to develop a Dillwyn soaprock body at Nantgarw—the porcelain recipes given by Dillwyn and Young are so different and there is a major problem since for both recipes the components are given in code! Dillwyn does provide a key to his shorthand coded script in his work notes, but Young does not, and Morton Nance makes an arbitrary assumption that the two coded lists are probably matched with each other. For example, Morton Nance suggests that Young used the following codes for his components: XV bone ash, LX potash, V sand and B china clay. The last two are identical to those used by Dillwyn but Dillwyn used the codes FX for potash and SR for soaprock—we may well ask where the bone ash is in Dillwyn's recipe and the soaprock is in Young's analogous recipe?

In this context, Morton Nance then suggested that perhaps Young was trying to emulate the Nantgarw recipe of Billingsley and Walker as their formulaic recipes are almost identical: we have seen that this is patently not so—indeed, Young had no idea of the formulaic Nantgarw recipe after the departure of Billingsley and Walker to Coalport in 1819/1820 and he certainly did not appreciate that bone ash was the major component quantitatively of their Nantgarw recipe.

What is clear from the summary statements cited above is the incredible confusion that has arisen over many years which has clouded the issue: stated simply, did Young make porcelain at Nantgarw and if so, what kind of porcelain was it? We can theorise that his experiments undertaken in 1819 and 1820 cannot have been commercially hugely successful in terms of output but his efforts may well have created a new ceramic at Nantgarw which was unlike the old soft paste porcelain, but perhaps the idea should not be as easily dismissed as previous authors have maintained?

William Weston Young's obsession with pottery and porcelain certainly extended over many years of his career from his earliest employment with Lewis Dillwyn decorating earthenware at the Swansea Pottery in 1803, until his final days in Bristol and the Midlands in the 1840s. Even after his departure from Nantgarw in 1823 as a disillusioned porcelain manufacturer the idea of marketing his version of "Nant-garw" porcelain was still fermenting. So, in 1838, Young returned to the theme of porcelain by re-invigorating his idea of issuing a *"ready-mix"* form of his "Nant-garw" porcelain for sale to porcelain and chain manufacturers. This idea was not as revolutionary as it may at first seem as it had been successfully launched in France in 1830 and had been hailed there as a useful business practice for French ceramics manufacturers. Young was then based in Bristol, where he had recently completed his book on the *Guide to the Beauties &c. of Glyn-Neath*, published with the sup-port of John Wright in 1835. Enlisting the help of John Wright at Bristol and John Lewis of Cheltenham, he installed machinery at Bristol for the grinding of porcelain materials for his *"ready-mix"* to sell to china manufacturers. The plan was to sell the porcelain frit in casks and then batches of china clay would be sold separately for addition in the requisite proportions to the consumer. This idea is a little perplexing at first as, of course, the preparation of porcelain frit does not merely involve the mixing of the required components but also needs some preliminary firing sequence to stabilise differential mineral hydrations and to remove organic residues which may be deleterious to the final mixture. Nothing is mentioned about this aspect of frit preparation but it was probably a *sine qua non* that some thermal treatment of the components was absolutely necessary: see, for example, the discussion in Edwards (*Nantgarw and Swansea Porcelains; An Analytical Perspective* 2018) where the frit preparation described for porcelain manufacture involved a high temperature cycle and precise grinding instructions before mixing with other components such as china clay, especially if hard minerals such as chert and flints were involved, as employed at Nantgarw under Billingsley and Walker. The underlying business reason for purchasing "*ready-mix*" frit was that it saved the porcelain manufacturer the machinery involved in the preparation, grinding, sieving and mixing of the frit components, outside of the time spent in their acquisition and transportation from individual sources. The hidden costs of transportation of raw materials was a significant burden in porcelain manufacture, particularly prior to the advent of railways, which only came in after the 1830 Rainhill Trials, and the only alternative then to lengthy road transport by waggons was canal and coastal transportation. The Welsh porcelain manufacturers at Swansea and Nantgarw sourced their materials from afar as quality was the overriding supply criterion: the best quartz sand and flints came from East Anglia, kaolin/china clay came from Cornwall, charcoal and coal fuel were derived from the best Welsh Rhondda Valleys anthracite steam coal. Other components, such as bone ash and glass cullet were normally not locally derived, and some porcelain manufacturers sourced their materials from overseas suppliers—such as charcoal from Saxony (Derby) and china clay from Virginia (Bow) (Anderson 2000; Ramsay and Ramsay 2007). So, it could be quite appealing regarding the time and effort saved in the preparatory stages for a ceramics manufacturer to purchase a good quality ready ground and treated frit to start their work.

In his later attempt to market the "Nantgarw *ready-mix*" frit, Young first made an approach to Apsley Pellatt of the London Pottery Co., a prestigious and well-respected company based in the capital, who were aware of the excellent quality and reputation of the Nantgarw china and of the financial appreciation with which it was accompanied some 18 years after the closure of the factory in 1820. In a letter to Apsley Pellatt on July 14th, 1838, Young proposed the sale of his Nantgarw *readymix,* claiming that only he was aware of the secret formulaic recipe (which of course he was not!) and that its quality was superior to and far exceeded any china being produced at that time in Staffordshire—most of which was of the bone china variety (Turner, *Ceramics of Swansea and Nantgarw* 1897). Young was fairly safe in this claim as William Billingsley had died in 1828 and the whereabouts of Samuel Walker at this time were completely unknown. Young was offering his "*ready-mix*" frit compound for sale at £15 per ton—it is not known if this included the transportation cost or not. However, Apsley Pellatt was not interested in Young's proposal, so Young then went to the Staffordshire ceramics factories and he did receive some tentative interest there from the china factories of Minton, Spode and possibly also Coalport. The eventual conclusion was that the potential purchasers expected the *ready-mix* porcelain frit after their own subsequent treatment, mixing and firing to produce the original Nantgarw quality porcelain body and it did not, but all agreed that it was nevertheless a quite acceptable and surprisingly good porcelain body. Despite this, it seems that no one wished to avail themselves of the purchase of this frit, which clearly had not reached their expectations for creating Nantgarw porcelain as they knew it, and Young's enterprise naturally collapsed forthwith.

In August 1838, John Lewis, Young's new partner in the "*ready-mix*" porcelain frit enterprise with John Wright, wished to inspect some finished specimens of Young's formulaic porcelain. Young seemed not to have any in his possession at that time but he managed to recover some plates that he had earlier sold to a local client in Neath and these then served as an example of his work: it is interesting that this is very strongly suggestive that Young had actually made porcelain according to his formula, albeit in a limited quantity perhaps, and had sold it commercially to several clients locally. We now know that he was unsuccessful in his wider marketing strategy for his new porcelain body frit—but he obviously had managed to make several specimens for local sale. The question remains, of course, as to where this porcelain was fired: we know that Young had become disillusioned following his trial experiments at the Nantgarw China Works in 1819–1821, but he records in several entries in his diaries made several years later that he had commenced porcelain trials again using the kilns at the Glamorgan Pottery in Swansea, then being run under the tenure of Baker, Bevan and Irwin, under an arrangement that he made with Mr. Bevan, one of the proprietors. So, it is very likely that the local porcelain items retrieved by Young and under discussion here could have been derived from a later batch fired at Swansea and made according to Young's formulation and recipe. It is intriguing, therefore, to postulate as to whether or not Young had impressed these items with his own version of the NANTGARW C.W. mark (Fig. 10.7)–namely, porcelain that had been formulated and possibly compounded at Nantgarw, by the owner of Nantgarw China Works, but decorated by him and fired elsewhere?

A further point of intrigue mentioned in passing by Morton Nance (*The Pottery and Porcelain of Swansea and Nantgarw* 1942, p. 407) relates to his ownership of a piece of porcelain called the "*Terrett*" mug (illustrated in Morton Nance, plates CLXXVII, C, D, E and F, facing page 407) which Morton Nance alleges was made by Young according to his Nantgarw formulation and given to John Wright as a birthday present for his younger brother, Terrett Allis Wright, sometime around 1838. Morton Nance comments that this piece, although it is a good and translucent porcelain, is nevertheless inferior to the Nantgarw porcelain of Billingsley and Walker! Surely, therefore, the suggestion that Young never made porcelain, either whilst at Nantgarw or after departing the Nantgarw China Works site, is untenable—but the question remains as to whether or not he ever made it commercially, outside of his experimental trials, whilst he was at Nantgarw even though if it was made elsewhere he seemingly marked it as Nantgarw anyway!

10.6 Comparison of Analytical Data

During this survey of the life of William Weston Young, it became apparent that his ceramics interests and accomplishments were not only diverse but have been open to some misinterpretation and conjecture. Although his achievement in the invention of the *Dinas* silica brick is never in doubt, the earliest historical accounts of his forays into the finer porcelain ceramics at Nantgarw are very much more subject to question. His liaison with William Billingsley and Samuel Walker in the establishment of the Nantgarw China Works during its phase I and II operations between 1814 and 1819 is never in doubt but there is a considerable difference of opinion concerning his claims and attempts to synthesise a Nantgarw-type porcelain at Nantgarw and elsewhere between 1820 and the 1840s. There are two schools of thought on this: namely, that William Weston Young did succeed to make porcelain at Nantgarw sometime after the departure of Billingsley and Walker in early 1820 and conversely, that he did not do so. Several authors of esteem and considerable historical standing, such as Jewitt and Turner (*The Ceramics of Swansea and Nantgarw* 1897), have maintained that Young did make porcelain at Nantgarw post-Billingsley and Walker, whereas others such as Eccles discredit that opinion. A third school can be identified who maintain that there must be some considerable doubt about this, such as Morton Nance, whilst others just do no mention it at all, such as Jenkins (*Glamorgan Historian* 1968) and Hillis (*Welsh Ceramics in Context* 2005). An obvious forensic criterion that would perhaps settle the matter is provided by an answer to the rhetorical question: if Young did succeed and achieve the manufacture of porcelain at Nantgarw or elsewhere, can it now be recognised and identified? This has in part been addressed and answered, if only ineffectively, by the response that Young had synthesised Nantgarw porcelain, which is so closely similar to that of the Billingsley and Walker formulation that it cannot therefore be discriminated and distinguished from the original! The answer to such a tenet must therefore lie in scientific examination and analysis, which has so frequently in the past contributed to the unmasking of fakes (Craddock, *Scientific Investigation of Copies, Fakes and Forgeries* 2009; Wieseman, *A Closer Look at Deceptions and Discoveries* 2010; Nickell, *Real or Fake: Studies in Authentication* 2009) and which must now have a future role to play in the attribution of unknown or suspect porcelain specimens to source factories (Edwards, *Nantgarw and Swansea Porcelains: An Analytical Perspective* 2018). Of course, if Young actually had used the precise Billingsley and Walker Nantgarw formulation for the synthesis of his ceramic body after they had departed Nantgarw then the detection of this as an interloper would indeed be deemed difficult by any analyst; as Orna (1996) has so correctly pointed out:

> Science can expose a fake but can rarely prove an object is genuine.

Forensically this is very true—but there are cases which are now well-documented where science has overturned expert opinion of an artwork as genuine or alternatively had categorised it as a fake and conversely, has caused a re-evaluation of a genuine original which now has to be catalogued as a fake or a copy.

10.7 Analysis of Nantgarw Shards

As stated above, Nantgarw porcelain shards and fragments of finished porcelain have been studied historically and reported in the scientific literature by Eccles and Rackham (*Analysed Specimens of English Porcelain* 1922), Tite and Bimson (1991), Owen et al. (1998) and by Owen and Morrison (1999). The sum total of 21 shards and 4 fragments of finished porcelain analysed historically have yielded the following results, further details being provided in Table 10.2: the distribution of analysed

Nantgarw shards is as follows, Eccles and Rackham (3 fragments of finished porcelain), Tite and Bimson (3 shards), Owen et al. (9 shards and 1 fragment of finished porcelain) and Owen and Morrison (9 shards and 1 piece of glass assumed to be potentially of Nantgarw glaze origin?).

All 3 specimens studied by Eccles and Rackham are classified as "normal" soft paste porcelain, as are the 3 specimens of Tite and Bimson. However, of the 18 shards studied by Owen et al. (1998) and Owen and Morrison (1999), only 14 were proclaimed to be "normal" and 5 were atypical of soft paste porcelain produced by the Nantgarw factory according to the Billingsley and Walker formulation as revealed later to John Taylor (*The Complete Practical Potter* 1847).

The 21 "normal" shards gave an average elemental oxide composition of; SiO_2 39–46%; Al_2O_3 12–14%; CaO 21–26%; P_2O_5 15–20%; K_2O 1–3% and Na_2O, MgO < 1%. These analyses agree very well with the compositional details of bone ash, sand and alkaline flux reproduced in the recipe for Nantgarw porcelain provided to Taylor by Samuel Walker.

The 4 saucer -and plate fragments of finished porcelain also gave analytical elemental compositional data which lie within these average values found for the "normal" shards. Nantgarw porcelain did not contain glass frit or cullet, so the presence of lead oxide would not be expected in the elemental oxide analyses of the fired body; however, Eccles and Rackham did note a trace of lead oxide in their analyses of the finished porcelain fragments, which can be attributed to their use of samples which incorporated both the body and the applied glaze (which does contain a significant percentage of lead oxide from the flint glass component) and the grinding preparation of these for the first stage of chemical digestion in the analysis would have therefore automatically included the glaze component along with the ceramic body. This conclusion is confirmed by the analytical result of Owen et al. (1998) on their glazed plate fragment, which when interrogated at a fractured edge free from glaze showed no evidence of a lead oxide signature in the body itself.

Hillis (*Welsh Ceramics in Context II* 2005, p. 178) compares the Nantgarw porcelain body composition with that of Derby porcelain and notes that Nantgarw contains significantly more potash alkaline flux and comments that both the Derby and Pinxton porcelain bodies show analytical evidence of a lead oxide component from the incorporation of glass cullet into their formulation recipes (Owen and Barkla 1997). Hillis then concludes that the Nantgarw body composition must have been inspired by Billingsley's work at the Derby China Works before he left for Pinxton in 1795. Yet, Derby porcelain contains glass cullet and Nantgarw does not—so William Billingsley must have made a conscious decision to exclude this component form his Nantgarw body paste for some reason.

Owen et al. (1998) have back-calculated a Nantgarw raw materials recipe based upon their analytical data, assuming that the bone ash is represented by apatite, $Ca_3(PO4)_2$, clay as kaolin, sand as quartz, SiO_2, and potash as pure potassium carbonate, K_2CO_3. The comparative percentage compositional data are as follows: Nantgarw recipe: Owen et al. data = bone ash 41.3: 42.0% (including calcite 43%), sand 22.2: 27.9%, potash 3.2: 3.0% and china clay 33.3: 27.1%. Bearing in mind the approximations made in terms of the purity of the raw materials used at Nantgarw

Table 10.2 Analyses of Nantgarw Shards: Porcelain Body/%

Specimen	SiO_2	Al_2O_3	CaO	P_2O_5	K_2O	Na_2O	MgO	Fe_2O_3	PbO	Analyst
NG1	58.1	19.5	10.5	8.5	2.6	0.1				E&C (2018)
NG2	49.1	16.7	12.1	18.5	1.4	1.5	0.5	0.2		E&C (2018)
NG3	53.5	16.6	10.7	16.6	1.4	0.6	0.2	0.4		E&C (2018)
NG4	68.1	20.5	4.1	4.0	1.9	0.9	0.1	0.4		E&C (2018)
NG5	44.0	14.8	17.6	19.8	1.9	0.8	0.9	0.2		E&C (2018)
NG6	80.1	12.6	0.3		1.2	2.6	3.1	0.1		E&C (2018)
NG7	65.5	17.6	6.4	7.7	0.6	1.6				E&C (2018)
NG8	47.5	17.1	12.8	19.3	1.3	1.2	0.5	0.3		E&C (2018)
NArt200	46.9	16.5	15.2	18.1	0.9	0.8		0.3	0.8	E&C (2018)
C586(22)	38.9	18.1	22.5	17.1	2.6	0.1	trace	trace		E&R (1922)
C510(21)	46.0	17.0	19.7	13.9	2.7	0.4	trace	trace		E&R (1922)
E3	40.2	20.9	22.0	13.1	1.7	0.9	0.7			E&R (1922)
N14-1	43.8	12.5	22.5	17.4	2.5	0.8	0.6			T&B (1991)
N18-4	45.0	13.3	21.2	16.4	2.2	1.0	0.5	0.4		T&B (1991)
N18-7	44.6	13.3	21.9	16.4	2.2	0.7	0.7	0.2		T&B (1991)
N1	43.9	12.5	23.0	17.3	2.3	0.4	0.4	0.2		O et al. (1998)
N2	70.8	8.9	9.9	7.5	2.3	0.2	0.2	0.2		O et al. (1998)
N3	43.9	12.3	22.8	17.5	2.5	0.4	0.5	0.2		O et al. (1998)
N4	80.3	9.1	0.6	0.5	5.6	1.8	1.9	0.1	trace	O et al. (1998)
N5	45.0	12.3	22.5	16.9	2.2	0.4	1.9	0.1		O et al. (1998)
N6	45.2	13.0	23.7	16.7	8.4	0.4	0.4	0.2		O et al. (1998)
N7	42.8	13.0	22.8	17.9	2.3	0.4	0.5	0.3		O et al. (1998)
N8	44.4	13.0	22.2	17.1	2.2	0.4	0.4	0.3		O et al. (1998)
N9	43.7	12.3	23.6	18.2	1.2	0.4	0.4	0.2		O et al. (1998)
N10	39.0	13.1	25.6	19.4	2.0	0.4	0.4	0.1		O et al. (1998)
N11	45.7	13.8	21.4	15.4	1.7	0.5	0.5	0.2	0.7	O&M (1999)
N12	46.1	14.0	22.2	16.4	1.7	0.5	0.6	0.2	0.6	O&M (1999)
N13	43.7	12.7	21.0	15.0	1.8	0.5	0.5	0.2	0.8	O&M (1999)
N14	42.4	13.4	23.7	17.4	1.9	0.4	0.4	0.2	0.2	O&M (1999)
N23	39.0	13.3	25.5	19.1	2.0	0.4	0.4	0.2	0.1	O&M (1999)
N24	41.1	13.3	24.3	17.8	2.3	0.5	0.4	0.2	0.1	O&M (1999)
N25	80.7	9.0	0.5	0.5	5.4	1.5	2.2	0.1	0.1	O&M (1999)
N34	79.0	6.6	0.6		12.2	1.6	1.0			O&M (1999)
N37	45.5	12.1	21.8	16.3	2.9	0.6	0.4	0.2	0.1	O&M (1999)
N43	80.5	9.3	0.4	0.4	5.4	1.5	2.2	0.1	0.2	O&M (1999)
N44	80.3	9.3	0.6	0.5	5.5	1.5	1.8	0.2	0.1	O&M (1999)

and the accuracies of weighing of the individual components used to make up the formulaic batch recipe each time then these data are remarkably close.

The analyses of the "abnormal" shards identified by Owen et al. (1998) and Owen and Morrison (1999), comprising 5 in number, and representing some 25% of the "normal" shards in number, lie outside the error bars and parameters of the data for the "normal" Nantgarw shards. The 5 shards, all of which were obtained from the Nantgarw China Works site, comprised 3 of biscuit porcelain (unglazed) and 2 glazed. All contain magnesia, MgO, and this unequivocally confirms that a soaprock component was included in their formulation. One particular shard (N2) possessed a low phosphorus pentoxide percentage which was less than half that of a "normal" shard and a very high silica component, whereas the other four shards had only a very small trace of phosphorus pentoxide and very high silica percentages of around 80%., twice that of the "normal" shards. These are significant analytical differences and an inevitable conclusion is that the "abnormal" shards contain soaprock as a raw material, which renders their analyses rather similar to that reported for glassy Swansea porcelain by Eccles and Rackham (*Analysed Specimens of English Porcelain* 1922)! It also brings the analytical figures close to those derived for the later trident porcelain of Dillwyn with its high soaprock and low bone ash content. In this context, it is rumoured that Dillwyn had prior knowledge of the Nantgarw formulation when he started up in the Swansea China Works with Billingsley and Walker in 1814/15, but he was concerned at the appallingly high kiln wastage problems encountered at Nantgarw which was attributed to the paste being too plastic upon firing in the kiln at high temperature and that this would also not be amenable for his Swansea porcelain production. The result was that he expressly instructed Walker to build a new kiln at Swansea to avoid the wastage—Dillwyn seems to be the first to reveal in his notes that this wastage at Nantgarw was approaching 90% and that it was a great concern for his operations at Swansea! It could be argued that Dillwyn would not have been so concerned at the potential failure of his impending Swansea porcelain body if it did not have a similar composition to that of its Nantgarw analogue—and this is borne out by Dillwyn's notebook comments on his experiments to vary the composition of the Swansea porcelain body, one of which he claims to be a close approximation to the Nantgarw analogue, and this contains soaprock!! This implicitly corroborates the idea that Dillwyn had already learned of a Nantgarw formulation early in Swansea's history and, furthermore, this recipe contained a soaprock component.

Equally clear is the corollary that when Billingsley and Walker started their Phase II Nantgarw operations in 1817 they had decided to dispense with the soaprock content and replace it with an increased percentage of bone ash: seemingly, in this case it follows that Young may have indeed been aware of the soaprock content in the putative Phase I Nantgarw formulation but he was not informed by Billingsley and Walker of their changes made to the Phase II body at Nantgarw. Speculative although this hypothesis is it does seem to fit the salient features and comments of the time: an alternative hypothesis can be promoted in that Billingsley and Walker introduced soaprock into their Phase II Nantgarw body in an attempt to harden their product but there is no evidence for this in the admittedly small numbers of perfect finished porcelain pieces that have been analysed hitherto. In summary, therefore, it is a possibility that the earliest Nantgarw porcelain body did perhaps contain a measured amount of soaprock, which would give the magnesia content recognised analytically in the shards examined, but the changes made to the later body removed this in favour of bone ash—and this correlates with the analytical determinations too.

Owen et al. (1998) report that they collected their shards which were "*scattered around the site since the location of the waste tip was unknown*"; this is a rather strange statement as Isaac Williams carried out an archaeological survey of the Nantgarw China Works site in 1932 and he discovered the waste tip in the form of a pit depicted in Fig. D.1, comprising several layers, the bottom layer of which was located about a metre below the current surface which contained only shards of Nantgarw porcelain and this was assigned to the Billingsley/Walker period from about 1817–1820. It can only be assumed that this waste tip location was lost in the interim until it was rediscovered a few years ago? The shards N2 and N4 were clearly silicaceous with a very high silica content of between 71 and 80% and a low phosphorus oxide content of between 1 and 7% and a low alumina content of about 9%, which should be compared with the analytical data for the phosphatic shards with a silica content of 43–45%, a phosphorus oxide content of 16–18% and an alumina content of 12–13%. An intermediate shard descriptor from the later work of Owen et al. (1999) gives percentages of silica of 71%, calcium oxide 10%, phosphorus oxide of 8% and alumina of 9%. The N4 shard was particularly anomalous in that it also contained a measurable 0.5% lead oxide which could only have originated from the addition of flint glass cullet—a practice which was carried out at the Swansea China Works but not at Nantgarw!

Owen et al. (1998) state in conclusion to their analytical work that,

> There is still no proof that the siliceous shards found amongst the ubiquitous phosphatic wasters at Nantgarw were actually manufactured at the site.

This statement really does not stand up to a forensic procedural study as there is no hint of any documentation supporting the idea that porcelain was bought into Nantgarw from outside to supplement their own production, whilst also still maintaining that there is equally no supportive evidence for the production of what appears to be hard paste porcelain at the site, at least until now. An interesting article by Jones and Jones (*Swansea & Nantgarw Porcelains* 1988) adds to the dilemma in a comment that was made that some non-archaeological Nantgarw porcelains appear rather "*glassy*" in texture and translucency—as one might expect for a hard paste variety! Owen and Morrison (1999) have reported that they have subjected several hundred Nantgarw shards, all presumed to be phosphatic, to deep ultra-violet radiation and all showed a clear reddish-brown coloured translucency with the exception of three which exhibited a fluorescent greenish-yellow coloured translucency characteristic of siliceous shards. They also theorise that the kiln that was specially constructed by Walker at Nantgarw in the period 1814–1817 was capable of attaining a temperature in excess of 1420 °C that would easily have been compatible with the firing of a hard paste porcelain body—and this would have cleared the way for Billingsley and Walker, if they wished, to replace the bone ash component in their formulaic recipe by increased quantities of soaprock and china clay.

In their earlier 1998 paper, Owen et al. comment that any secular variation in the Nantgarw body composition cannot be evaluated from their shard specimens as these had lost their archaeological depositional context—the importance of this criterion is seen for Isaac Williams' burial pit excavation shown in cross-section here in Fig. D.1. The critical importance of logging the exact position of undisturbed wasters in the burial pit is thereby enhanced for subsequent analytical investigation and its interpretation—this procedure would facilitate a proper correlation to be accomplished between the analytical data and the location of the waster in its burial site and perhaps would go a long way to answering the question as to whether or not Billingsley and Walker did undertake trial experiments at Nantgarw, which were not recorded for posterity, unlike Dillwyn's similar exercises which were carried out at Swansea during 1815–1817. At Swansea, Dillwyn empirically and logically changed his paste body, starting with a phosphatic porcelain and depleting the clay and silica components from 44 to 27% and 52 to 20%, respectively; he realised that thereby the alumina: calcium oxide ratio was the primary factor in increasing the plasticity of the paste and in reducing the kiln wastage (Rado, *An Introduction to the Technology of Pottery* 1969). The composition of bone china and the effect of other components in the paste upon kiln firing at high temperatures and porcelain plasticity has been explored in detail in a series of detailed research papers by St Pierre (1954, 1955, 1956).

It is interesting to consider the shards studied by Owen and Morrison (1999) in greater detail:

N11–N14: sagged phosphatic shards, from cups, all lead glazed, where the edges were curled back upon themselves as a consequence of "overfiring" or of keeping too long in the kiln at too high a temperature. Analytical data averaged as follows: SiO_2 42–46%, Al_2O_3 13–14%, CaO 21–24%, P_2O_5 15–18%, K_2O 1–2%, Na_2O, MgO Fe_2O_3 < 0.5%. The conclusion is that the absence of magnesia indicated a zero soaprock component, and a low iron oxide presence is indicative of the use of a very pure quartz sand.

N23, N24: shards both impressed with the Nantgarw mark, with sagged edges, from the central parts of plates. Analytical data averaged as follows: compared with the other phosphatic shards (N11–N14) these showed a slight depletion in silica at 40%, and a similar bone ash component at 16–18%, other components the same. These small differences could reflect a variation in recipe but more likely are a result of possible serious cumulative errors in the weighing of the raw materials for making up the porcelain paste—for a detailed discussion of this aspect and the possible effect on analytical data the reader is referred to Edwards (*Nantgarw and Swansea Porcelains; An Analytical Perspective* 2018).

N25, N43, N44: fragments of cups and saucers, all glazed, which analysed as siliceous shards—all three demonstrated a greenish-yellow translucency to ultra-violet radiation.

N34; a shard of clear glass, initially assumed to be frit or cullet used by Billingsley in his composite. However, as Billingsley did not incorporate glass into his body paste we should conclude this shard is an interloper or perhaps may have been incorporated as a raw material in the Billingsley glaze.

N37: a successfully fired plate, impressed with the Nantgarw mark, moulded with the characteristic floral embossed border with ribbons and decorated with overglaze enamels. Analysis reveals a slightly higher silica content of 46%, compared with that associated with the normal phosphatic body range, but errors in weighing or determination of elemental oxides could have contributed to this.

Tite and Bimson (1991) specifically address the errors in elemental oxide determination in their SEM/EDAXS analyses and suggest these are ± 3% for the major components and significantly more for the others. Colomban (personal communication, 2018) has suggested that the errors from similar analyses could be as high as 5% or more for the major component determinations and 25–100% for the minor components and traces. A major factor in the comparison between the shard data from different determinations is the nature of the analytical study being carried out: the analytical technique is microscopic and depends upon the homogeneity of the specimen. Although we refer to the porcelain paste as a "melt" at high temperatures, in fact it is very domain dependent, and this is affected by the particle size and the diffusion of chemical species between domains after reaction and this is not the same as solution studies. Hence, the SEM/EDAXS experiments are usually carried out over multiple sampling positions in the specimen and involve sub-micron specimen interrogation, so these results often give a different analytical picture to the bulk methods, such as wet chemical digestion, although they do not involve such a great sample consumption. Owen et al. (1998) and Owen and Morrison (1999) do not present analytical errors in their elemental oxide determinations and this means that their data could sensibly and realistically overlap with those of Tite and Bimson and the bulk determinations of Eccles and Rackham. For example, the data of Tite and Bimson and those of Eccles and Rackham actually agree to within 3% for silica, 15% for alumina and lime, and 25% for magnesia, which are all within the confines of the determinational error defined by Colomban (P. Colomban, personal communication to the author). The Tite and Bimson data for their three Nantgarw shard specimens agree well with each other; the bone ash raw material component of the initial paste calculated on the basis of it being composed of hydroxyapatite, $Ca_5(OH)(PO_4)_3$, is 43% by the summation of the determinations for lime, phosphorus pentoxide and silica (which is calculated as total silica minus the clay component), which is in excellent agreement with the reported formulaic recipe of John Taylor (*The Complete Practical Potter* 1847). They also found the high temperature products of reaction, namely anorthite $CaAl_2Si_2O_8$, and whitlockite $Ca_3(PO4)_2$, which are present in the glassy matrix of the Nantgarw body in three ranges of particle sizes, namely, *large* at 20–40 microns, *average* at 5–10 microns and *very fine* with less than 1 micron diameter.

Eccles and Rackham assumed Dillwyn's comment of "*a beautiful china which stands well but rather too soft for the hard glaze*" in his work notes regarding his experimental trials of alternative porcelain bodies conducted at Swansea with Walker between 1815 and 1817 actually referred to the *Nantgarw body*. If this were true then this would mean that the composition of this same body would be: 9 parts china stone (feldspar), 12 parts china clay (kaolin), 12 parts bone ash and 3 parts lime. This would indeed be a curious composition for the accepted Nantgarw composition as we have come to accept today – but perhaps it does reflect again on the idea that Dillwyn had received another formulaic version of their Nantgarw paste given to him by Billingsley and Walker and this might even be the trial Nantgarw paste that they had decided to use in the Phase I operation at the set-up of the Nantgarw China Works in 1814. In that case, the modification made for the Phase II body would not have been too great in that the exclusion of the feldspar and an increased compositional component of bone ash would have been all that was required! Nevertheless, several percentage figures and ratios expressed by Eccles and Rackham do not add up: for instance, they suggest that this recipe equates with a ratio of alkali: alumina = china stone: china clay = 1:4, which does not match at all with the Nantgarw recipe. Also, a curious estimation of the bone ash component was derived from a summation of the lime plus phosphorus pentoxide analytically determined values only, and this does not account for the lime added (25% of the bone ash!), and the "alkali component" was the summation of the determined soda and potash only, neglecting the significant lime component used in the recipe provided.

10.8   Combination of Elemental and Molecular Analytical Techniques

Recent analyses of Nantgarw porcelain have adopted a combined approach using elemental and molecular spectroscopic techniques, namely, SEM/EDAXS to determine the relative compositions of elements and elemental oxides in the body, pigment and glaze formulations in Nantgarw shards and Raman spectroscopy to determine the molecular species of the minerals and their derivatives involved in the body, pigments and glaze in the shards.

The analytical approach undertaken has been as follows:

Examination of the elemental composition of the body, glaze and any applied pigment where appropriate of selected shards excavated from the Nantgarw China Works site using the SEM/EDAXS technique.

A parallel study of the molecular composition using Raman spectroscopy of the Nantgarw body in shards from the same source batch and of perfect specimens of Nantgarw porcelain from a private collection which includes some rare and important examples dating from the period 1817–1819 (Edwards, *Swansea and Nantgarw Porcelain: A Scientific Reappraisal* 2017) (Appendix D). This study facilitated the analytical characterisation of the Billingsley/Walker/Young Nantgarw body formulation from the shards demonstrated the viability of the technique for establishing the acquisition of body data through the superficial glaze without necessitating its removal and thereby effecting a comparison with the perfect porcelain examples, some of which were marked with an impressed NANT-GARW C.W., and others unmarked. This meant that for the first time it became possible to verify scientifically the attribution of unmarked Nantgarw porcelain pieces.

The analytical information acquired from the application of both techniques in this combined approach is very different: the SEM/EDAXS experiment will provide quantitative details about the silicon, aluminium, magnesium, calcium, sodium, potassium iron and phosphorus content of the body and glaze along with lead, which can be related back to the component raw materials such as quartz, sand, clay, soaprock, flint glass cullet, lime and potash, whereas the Raman spectroscopic experiment identified the chemical compounds that are formed at the high temperatures in the firing or glost kilns, such as wollastonite, mullite, anorthite, whitlockite, cristobalite and enstatite.

10.9 SEM/EDAXS Analyses of Nantgarw Shards

The batch of shards studied using the combined approach techniques were provided from the 1995 excavation of the Nantgarw China Works site carried out jointly by the Rhondda-Cynon Taff Archaeological Trust and the Glamorgan Archaeological Society (Murphy 1997). These shards comprised biscuit porcelain, glazed fragments and three enamelled fragments; the latter were most unusual and had not been described as being found at the site hitherto. Body compositional data for shards NG1-NG8 are presented in Table 10.2 along with others that have been reported in the literature form Eccles and Rackham (1922), Tite and Bimsom (1991), Owen et al. (1998) and Owen and Morrison (1999). A further shard, NArt200, is unique because it is a biscuit porcelain shard from a specially introduced porcelain body manufactured in 2017/2018 according to the original Nantgarw recipe to celebrate the 200th anniversary of foundation of the Nantgarw China Works in full production. It will therefore be interesting to compare the analytical, elemental and molecular spectral signatures of the new Nart200 shard with the analyses of the original shards from the site.

Compositional data for Nantgarw shards falls into three broad classes: "normal" phosphatic Nantgarw porcelain which matches the data from analyses performed by Eccles and Rackham (1922), Tite and Bimson (1991), Owen et al. (1998) and Owen and Morrison (1999), "intermediate" or hybrid porcelain (such as NG4 and NG7 reported here) and "abnormal" siliceous porcelain (represented by shard NG6 here). The detailed percentages of the elemental oxides found in the shards NG1-NG8 and the NArt200 shard are given in Table 10.2 and corresponding the glaze compositional data where determined are given in Table 10.3 along with the analogous data from previous shard determinations reported in the scientific literature.

Table 10.3 Analyses of Nantgarw Shards: Glaze/%

Specimen	SiO$_2$	Al$_2$O$_3$	CaO	P$_2$O$_5$	K$_2$O	Na$_2$O	MgO	Fe$_2$O$_3$	PbO	Analyst
NG2	76.7	9.7	9.0		1.2	2.3	0.3	0.2	0.5	E&C (2018)
NG6	84.5	9.6	0.2		1.1	2.2	1.1	0.5	0.8	E&C (2018)
NG7	75.7	13.5	6.3	0.2	1.8	1.0		0.3	1.2	E&C (2018)
NG8	66.9	18.7	5.1		1.5	3.8	2.1	0.3	1.6	E&C (2018)
N14-1	60.0	12.9	11.0	0.3	3.3	0.7	0.3	0.4	11.1	T&B (1991)
N18-4	61.6	11.3	8.8	0.8	2.3	0.7	0.4	0.5	12.8	T&B (1991)
N18-7	62.6	11.4	10.4	1.2	2.1	0.7	0.3		11.3	T&B (1991)
N25	67.8	7.0	1.1	0.4	4.3	1.1	2.9		15.4	O&M (1999)
N37	53.4	9.3	11.6	0.3	2.3	0.5			22.6	O&M (1999)
N43	66.0	6.3	0.9	0.3	6.1	2.7	1.6		14.7	O&M (1999)
N44	68.3	7.4	0.9		5.2	1.3	2.6		14.4	O&M (1999)

The average elemental oxides/% found in the body compositions from our results within these three categories are:

Shard(s)	Category	SiO$_2$	Al$_2$O$_3$	CaO	P$_2$O$_5$	K$_2$O	Na$_2$O	MgO	Fe$_2$O$_3$	PbO
Body composition										
NG1,2,3,5,8	Normal	52	17	12	15	2	1	<1	<1	0
NG4,7	Intermediate	67	19	5	6	1	1	0	<1	0
NG6	Siliceous	80	13	<1	0	1	3	3	0	0
Glaze composition										
NG2,8	Normal	72	14	7	0	1	3	1	<1	1
NG7	Intermediate	76	14	6	<1	2	1	0	<1	1
NG6	Siliceous	85	10	<1	0	1	2	1	<1	1

In comparison with the detailed data presented by Owen and Morrison (1999), the Edwards and Colomban recent data on the "normal" Nantgarw phosphatic body show a higher silica and alumina content by several %, a significantly smaller lime content by 10% and identical phosphorus pentoxide, potash and soda percentages. With the possible exception of the lime percentage value, these differences in the two data sets can be explained by a summation of thee experimental errors in the elemental oxide determinations, which can be estimated (Tite and Bimson 1991) to amount to at least 5% or more for the major components and perhaps as much as 25% for the minor components.

The "intermediate" shards identified here, namely NG4 and NG7, are not specifically discussed elsewhere but their presence along with the "normal" phosphatic bodied shards at the Nantgarw China Works site indicate that at some stage trial experiments on body composition variation must have been carried out there, as Lewis Dillwyn did at the Swansea China Works between 1815 and 1817. What we cannot ascertain is when this occurred or who was responsible for it? Did Billingsley and Walker undertake such experiments in an attempt to improve the very high kiln wastage suffered by their Nantgarw phosphatic body or did these two shards emanate from the attempt by William Weston Young to re-create Nantgarw porcelain? The former explanation is a better concept because from his written records Young never realised that bone ash was a component of the Billingsley and Walker recipe—and these shards comprise a significant percentage of phosphorus pentoxide, albeit 6% in place of the "normal" 15%, and a much smaller lime content of 5% compared with 12%.

The "abnormal" siliceous shard NG6 best describes the hard paste porcelain formulation of William Weston Young and is so far removed from the predominantly phosphatic body recipe of Billingsley and Walker that it would be difficult to envisage any progressive variation on their part that would have resulted in such a completely different outcome compared with their "normal" formulation. It is significantly higher in silica content and much lower in lime content than the "normal" phosphatic shards of Billingsley and Walker – but the really important result is the percentage of phosphorus pentoxide in NG6, which is zero! Also, we see for the first time a percentage of magnesia amounting to 3%, which correlates with the addition of up to 12% soaprock in the initial formulation. This is closely reminiscent of the recorded trial experiments undertaken by Young and a reasonable conclusion is that this shard is evidence that Young did make porcelain at Nantgarw but of the hard paste type category. Examination of the data of Owen and Morrison (1999) indicates that they too found a siliceous, non-phosphatic, shard in their batch of Nantgarw material, whose silica and alumina percentages differ from those of NG6 quoted here by several %, but still within the acceptable cumulative experimental errors cited above. Clearly, some form of experimentation had been advanced at the Nantgarw China works site in the preparation of hard paste porcelain. We should also refer to the period in the 1850s, when the Nantgarw China Works was under the tenure of William Henry Pardoe, who advertised that he had commenced the manufacture of porcelain there again—but it must be recognised that if this shard were part of Pardoe's production than it would be expected to be mixed up with his earthenware shard output, which was more considerable and also found at higher levels in the waste pit (see Fig. 10.3, and Williams 1932).

The various data from different batches of shards and fragments support each other favourably and reinforce the hypothesis that some experimentation in the composition of the porcelain body had been undertaken at Nantgarw in contrast with Herbert Eccles' statement that the Nantgarw body was "*invariant*" from his inspection of Nantgarw porcelain on show at the Glynn Vivian Gallery Loan Exhibition, Centenary Celebration, 1914. What is not clear, of course, is the extent to which the commercial production of any trial bodies was put in place; it is appropriate to consider a potential scenario and a chronology under which such trial experiments and production might have occurred. Firstly, could William Billingsley and Samuel Walker have created a "hybrid" version of their Nantgarw formulaic body which contained a measureable amount of soaprock which would give the analytical signature for magnesia found in several of the shards, which also contained a significantly smaller amount of phosphorus pentoxide, indicating a reduction in bone ash in the recipe? It may be suspected that little Nantgarw porcelain would have been made in the latter part of 1814 when experiments were under way to test the new kilns with small batches under different firing conditions, culminating in the special samples made to accompany the *Memorial* submission in September of that year. It is sad that the order books detailing commissioned pieces and sales do not exist for the Nantgarw China Works as these would have provided a source of data for an estimation of the production figures at that time. When operating at full production other porcelain factories (for example, Pinxton, from a letter of William Billingsley to John Coke in 1798) have noted that each kiln could take a batch firing of up to some 25 dozen pieces, which would require approximately a week for the heating, constant temperature and cooling cycles to be maintained, but it is unlikely that Nantgarw would have been operating at anything like this level at the start-up of phase I in 1814 bearing in mind the paucity of financial backing from William Weston Young for the running costs and the purchase of raw materials, gold for gilding and the pigments for enamelling. After the failure to secure the requisite government support for their continued operations at Nantgarw, Billingsley and Walker rapidly joined forces with Dillwyn at Swansea, within a time frame of about three weeks, so there would have been little opportunity to undertake further experimentation at this time. No mention is made of stock dispersal prior to or immediately following Billingsley and Walker's transfer to Swansea s it may be concluded that very little china was in existence at Nantgarw.... but did they take some to Swansea for decoration or perhaps use the moulds there later? Secondly, did Billingsley and Walker apprise Dillwyn of their formulaic recipe to kick-start the china manufacture operations at Swansea? Eccles assumes that they did, and that somehow Flight, Barr and Barr got wind of this and sent their cautionary and rather threatening letter sequentially to Walker and Dillwyn at the Swansea China Works as they may have appreciated the impending danger to themselves of the association of personnel there. It is interesting that Flight, Barr and Barr did not address a copy of their letter to Billingsley and does this imply that perhaps they regarded Walker as the mastermind behind the actual china production operations and Billingsley as the artistic decorator? Thirdly, it seems unlikely that having achieved the phenomenal success with their phosphatic body in the London salerooms in 1817 at the start-up of phase II of the Nantgarw China Works, that Billingsley, Walker and Young would then revert to trying out a different recipe when the accent and pressure would be on the creation of as much porcelain of this successful body composition as they could to satisfy the almost insatiable demand of their clientele. Put in another sense—why would they have tried to make a "better version" of a successful product when the demand was there: around this time Dillwyn was trying to improve the robustness of his celebrated duck-egg porcelain body at Swansea by incorporating soaprock at the expense of china clay and bone ash. He eventually produced his trident body, with a much more robust body, but immediately lost his market edge in London and locally because of the inferior translucency resulting in the rapid demise of the Swansea China Works. This message would possibly not have been lost upon Billingsley and Walker at Nantgarw and would have provided a grim warning of undertaking such "improvements" to an already successful product.

10.10   Nantgarw Shard Glaze Analyses

The composition of the original Nantgarw glaze has entered much discussion since its recipe has been lost: although Samuel Walker is alleged to have provided John Taylor with the details of the body paste recipe for Nantgarw biscuit porcelain which was published in his book in 1847 (*The Complete Practical Potter* 1847), and again published in the *Pottery Gazette* of 1885, no information was forthcoming about the composition of the glaze used by William Billingsley and Samuel Walker between 1817 and 1820—the so-called No. 1 glaze, or Billingsley glaze. William Weston Young and Thomas Pardoe invented a glaze to prepare the remnant Nantgarw stock for sale from 1820–1823 and this is referred to as the No. 2 glaze: Young recorded the composition of this glaze and it was used by Professor Mellor in the 1880s to evaluate the synthetic porcelain made according to the Nantgarw formulation and he pronounced it satisfactory for the purpose and adheres well to the biscuit body—the No. 2 glaze is not the same as the No. 1 glaze as it has a creamier translucency and is commonly referred to as the Pardoe glaze as it invariably accompanies his enamelling decoration on Nantgarw porcelain. The Young/Pardoe glaze contains many more ingredients and additives that the porcelain body and in the glost kiln, which is fired to a lower temperature than the main biscuit porcelain kiln, several reactions occur to transform the composite minerals into a clear translucent and protective coating. These are reproduced along with others which are created in the biscuit porcelain body in Table 2.2, such as wollastonite, diopside, whitlockite, mullite and anorthite (Bimson 1969); the essential requirements for a glaze in addition to its transparency are a compatibility with the subsurface biscuit porcelain body in terms of a similar thermal expansion coefficient, which minimises the fracture or differential cracking between the glaze and the body, and an ability to preferentially absorb the mineral enamels used in the overglaze decoration to give an adherence fastness and a depth of colour to the resultant decoration. Generally, after firing of the enamelled decoration, a further glazing coating is applied to protect the enamelling from damage during use and cleaning of the items.

   The glazes on porcelain can be classified into three types: low lead, containing between 10 and 15% lead oxide, medium lead, containing between 30 and 40% lead oxide, and high lead, containing higher than 40% lead oxide. In 1820, John Rose of the Coalport China Works was awarded a medal for inventing a lead-free glaze for porcelain as by that time the cumulative toxic dangers of the use of lead and exposure to its compounds in ceramics factories were well appreciated. The lead was usually added to a glaze recipe in the form of flint glass, which is not to be confused with flints, which were chemically totally different in that they were almost pure silica, SiO_2. Flint glass could have a variable composition (Edwards, *Nantgarw and Swansea Porcelains: An Analytical Perspective* 2018) in preparations from between 10 and 60 + % lead oxide in molten silica and a good working standardisation figure for ceramics analysts is that the percentage of added cullet (finely ground flint glass) in the glaze recipe could be regarded as about 4X the lead oxide determination. For example, 5% lead oxide would indicate a cullet percentage of about 20% in the recipe, but obviously this would be very dependent upon the type of flint glass used in the process. In some cases, the flint glass was replaced by crown glass (or soda glass) cullet of a lower refractive index than the analogous flint glass, and occasionally the lead oxide was replaced by tin oxide (or a tin carbonate) of lower toxicity (*majolica* earthenwares, for example, are tin-glazed and not lead-glazed). Generally, much fewer analyses have been carried out on the glazed components of Nantgarw shards than on the porcelain bodies and some important conclusions derived from these analytical data are given below:

Tite and Bimson (1991) present full analytical data for three Nantgarw shards (reference numbers from the British Museum archive: 32,846, 32,847 and 32,714) as follows: SiO_2 60–63%, PbO 11–13%, Al_2O_3 11–13%, Na_2O ~1%, K_2O 2–3%, MgO < 0.5% CaO 9–11%, Fe_2O_3 < 0.5% and P_2O_5 0.5–2%. It can be concluded therefore that the three phosphatic porcelain shards have a low-lead glaze (whose percentage lies in the range 10–15% lead oxide). Also sample 32,714, coded N18-7, is significantly higher in phosphorus pentoxide content compared with the glazes on the other two shards. The phosphorus would have been added in the form of bone ash, hydroxyapatite $Ca_5(OH)(PO_4)_3$. In all three shards the magnesia value is very low and it can be inferred that it originated from a small amount of soapstone additive, probably talc $Mg_3Si_4O_{10}(OH)_2$.

Owen and Morrison (1999) also report analytical data for glazes from three Nantgarw shards, namely, N25, N43 and N44. In contrast to the results given above by Tite and Bimson for their "normal" phosphatic shards, Owen and Morrison's data all refer to the "abnormal" classification, i.e. siliceous shards, and found: SiO_2 66–68%, Al_2O_3 6–9%, Fe_2O_3 0%, MgO 1–3%, CaO ~1%, Na_2O 0.5–3%, K_2O 4–6%, P_2O_5 < 0.5% and PbO 14–15%. This implies again that the glaze is still low in lead as used for the phosphatic porcelain body; otherwise the elemental oxide components are all very similar indeed except perhaps for the increased magnesia and decreased lime contents. Owen and Morrison also reported the results from their N37 specimen, a finished, glazed and decorated Nantgarw plate whose body was phosphatic in type. The results are as follows: SiO_2 53%, Al_2O_3 9%, Fe_2O_3 0%, MgO 0%, CaO 11%, Na_2O 0.5%, K_2O 2%, P_2O_5 < 0.5% and PbO 23%. Despite a significant increase in lead content of about 50% more than the siliceous shards, this phosphatic glaze is still deemed to be low in lead as it only contains half as much lead oxide as that required for a medium lead glaze. There is no soaprock additive and the bone ash content is still very low, but the silica is less with significantly more lime as an additive flux: all other components are almost identical for the two types of porcelain body used, which completely refutes the suggestions made in earlier literature that siliceous porcelain requires a different glaze to its phosphatic analogue because a phosphatic glaze would be incompatible unless it had a larger percentage of soaprock and silica added. Obviously, the formation of sagged porcelain in the biscuit kiln through over firing at high temperatures was a major problem that needed addressing by Billingsley and Walker but there is now evidence in existing records or historical documentation that they decided to undertake experiments to materially change their body composition as a result, which would have involved time-consuming trials, the provision of additional resources for raw materials and increased costs for kiln fuelling with little or no output porcelain being made for sale.

The SEM/EDAXS analytical data of Edwards and Colomban (2019) for the Nantgarw shard glaze compositions are very similar to those reported by Owen and Morrison (1998) and by Tite and Bimson (1991) but with a lower lead oxide content (1% in agreement with Owen and Morrison but significantly lower than Tite and Bimson at 10%)– again the significant detection of magnesia implies a soaprock component. The glaze composition from Edwards and Colomban is very closely similar for both the siliceous and phosphatic species shards—verifying the comment of Dillwyn that his soaprock siliceous porcelain body required a magnesia content in the glaze for better body adhesion.

Fourthly, the final scenario moves to the activities of William Weston Young's tenure of the Nantgarw China Works from 1820 until its closure in 1823. In his diaries we can see evidence of his attempts at undertaking porcelain manufacture there and the variations he made in the formulaic recipes: it appears that his compositions are exactly what we would have expected from the latest analyses of the Nantgarw shards, especially the siliceous shard specimen, NG6—which is high in silica, and which contains magnesia (from a soaprock additive), low in phosphorus and contains no flint glass cullet. It is reasonable to assign the NG6 shard, therefore, to William Weston Young's experimental porcelain manufacture at Nantgarw.

Fifthly, there is evidence that William Henry Pardoe also made porcelain at Nantgarw of unknown composition, which he sold as "Nantgarw china", following on from his successful earthenware manufacture and clay pipes business. It is postulated that he may have attempted to fuse the two businesses and to have created a glassy coating for his earthenware's using china clay and flint glass, so making a type of "pearl ware" which was often mistakenly confused with porcelain—it is believed this is what happened at the Cambrian Pottery prior to Lewis Dillwyn's establishing the Swansea China Works. The shards excavated from the waste pit and discussed here are truly porcelain and not a coated earthenware, as are the others enumerated in the Tables and studied earlier, and Pardoe's shards are evidently seen at higher levels in the waste pit as determined by Isaac William's excavation in 1932. A search for porcelain-like shards from Pardoe's work should be located in Level 3 of the waste tip along with clay pipe shards and other earthenware materials.

If we then assume correctly that the siliceous shards found at the Nantgarw China Works originate from Young's experimental trials in the manufacture of a Nantgarw porcelain the analyses have produced a very important result which affects the historical appreciation of porcelain manufacture there: these shards would have been located at the upper part of level 5 in the waste tip and further excavations might be required to expose more of these specimens. Also, a parallel search to locate finished specimens of Young's porcelain should be undertaken to verify spectroscopically that although Nantgarw - like, and possibly even marked as such as has been suggested, they are truly hard paste porcelain versions.

10.11   Nantgarw Shards: Pigment Analyses

No analytical data have previously been reported for the pigments used in the decoration of Nantgarw porcelains: this perhaps not surprising in that very few pigmented or decorated shards have been discovered. Isaac Williams commented on this in his first excavation of the Nantgarw China Works site in 1932, when from some hundreds of biscuit and glazed shards collected from the waste pit he found none with any applied decoration. This could well be indicative that most of the wasters found in the pit came from the biscuit porcelain firing and first glaze application and that the china that survived these operations without sagging warping or exhibiting firing cracks would have been the most likely to have been decorated and sold on, without further damage being incurred. However, as part of a consignment of shards from the Nantgarw China Works Museum amounting to several kilos in weight and sent to the author personally, two small shards showed evidence of an applied pigment decoration as well as glazing and as such these represent a unique opportunity for analytical investigation in an attempt to determine the pigments used by William Billingsley or Thomas Pardoe to decorate Nantgarw porcelain locally and for which we have no records currently available as to the minerals they utilised for this purpose. Enamels on porcelain and fired ceramics are technically very different from those used in other artworks, oil paintings, frescoes and wall paintings since the high temperatures required for the firing or over glazing processes are not generally conducive for the retention of colour in many cases: for example, common pigments used in ancient times such as cinnabar(vermilion), azurite, malachite, verdigris, goethite, realgar, cerussite and orpiment are decomposed in air at elevated temperatures into metal oxides which have very different colours from the original minerals. Particularly sensitive are minerals which are hydrated, and composed of carbonates, sulfates and nitrates, such as covellite, cerussite, gypsum and calcite. In the literature, an account is given of the modern restoration of a 17th century fresco by Antonio Palomino in the Church of Sant Joan del Mercat in Valencia which was destroyed by gunfire and conflagration during the Spanish Civil War in 1939 and the temperatures generated (believed to be in excess of 600 °C) have converted most of the pigments such as blue azurite into brown copper (II) oxide (Domenech-Carbo et al. 2012). Modern pigments such as azo-dyes and phthalocyanines have a range of hues which are much appreciated by artists but being synthetic organic chemical compounds they also have a low tolerance to heat and would be unusable for ceramic work unless a cold glazing process was adopted. There is no information available whatsoever concerning the pigments and enamels that were used at Nantgarw by Billingsley, Young and Pardoe, and even Young's diaries, which usually provide a plethora of facts about his casual purchases, fail to make any mention at all about the acquisition of pigments for the decoration of Nantgarw porcelain. It is possible that Young and Pardoe used the enamels left over from Billingsley's artistic work when he left Nantgarw but these were quite expensive materials and it is unlikely that sufficient quantities would remain in stock to enable them to carry on for another two or three years. The only purchase of this sort which is recorded in Young's diaries is that of gold—which everyone has assumed naturally is required for gilding the finished china, but a *non sequitur* follows in a statement made by Young and Pardoe that gilding of their china was too deemed to be expensive and they sought to replace gilding with enamelling!

For our analyses, two different analytical techniques were employed, namely, SEM/EDAXS for the elemental oxide quantitative determinations as has been used previously by Tite and Bimson (1991), Owen et al. (1998) and Owen and Morrison (1999), and for the first time Raman spectroscopy for a molecular spectroscopic identification of the chemical moieties involved. Two pigmented, enamelled shards were selected for analysis: NG8 and NG9, the former containing evidence of a deep purple pigment and gilding, and the latter deep purple, blue, maroon and green pigments and gilding. The results for the quantitative elemental oxide determinations are shown in Table 10.4.

Raman spectroscopy (Long, R*aman Spectroscopy* 2002) has already established a demonstrable analytical success particularly for the characterisation of pigments used in the historiation of mediaeval manuscripts and a comprehensive database of the key signatures of mineral pigments exists (Bell et al. 1997; Burgio and Clark 2001). Application to fired ceramics provides another data source which informs the present investigation (Edwards et al. 2004; Edwards 2015; Kock and de Waal 2007; Colomban et al. 2001, 2004a, b; Colomban and Milande 2006; Ricciardi et al. 2009; Colomban 2013; Carter et al. 2017). Raman spectroscopy of the pigments used on the Nantgarw porcelain shards reveals some interesting molecular features which in combination with the elemental oxide determinations provide an indication of the minerals employed in the enamelling decoration of Nantgarw porcelain: to the author's knowledge this has not been reported hitherto:

Blue enamel: NG9 shows a strong peak at 823 cm^{-1} with a shoulder at 785 cm^{-1} characteristic of lacunar apatite, a complex sodium potassium cobalt and lead arsenate of formula $Na_{0.4}K_{0.1}Co_{0.5}Pb_4(AsO_4)_3$, which is quite usually seen in 18th Century enamels made from arsenic-rich "cobalt blue". Both Co and As, cobalt and arsenic, are determined by the SEM/EDAXS technique. Also, the characteristic D and G doublet bands of sp^2 and sp^3 hybridised carbon at 1320 and 1580 cm^{-1} are seen in the spectrum which is indicative of the addition of powdered amorphous carbon (usually in the form of charcoal, soot, or lamp black) which has been used to darken the colour of the pigment. Silicate bands near 985 and 1020 cm^{-1} are attributed to a lead glass additive to the pigment.

Maroon: NG9 shows the presence of haematite, iron (III) oxide Fe_2O_3, with bands at 220 and 285 cm^{-1} and again the addition of carbon as a darkening additive.

Green: NG9 shows the presence of Naples yellow, pyrochlore $PbSn_xSb_{2-x}O_7$, with peaks at 138, 452 and 505 cm^{-1}—a tin-rich enamel. Carbon is again observed as a darkening agent. The yellow pigment would have been mixed with a blue pigment,

Table 10.4 Analyses of Nantgarw Shards[*] : Enamelled Pigments/%

Specimen/colour	SiO_2	Al_2O_3	CaO	P_2O_5	K_2O	Na_2O	MgO	Fe_2O_3	PbO	CoO	SnO	Au	Other
NG9/maroon	67.5	8.8	3.0	0.4	2.0	1.9	2.2	13.2	0.7		0.2		
NG9/green	77.6	12.3	0.8		0.6	4.1	3.7	0.2	0.2				Cu 0.3
NG9/purple	86.6	3.4	0.6		1.2	3.3	1.3	0.3	1.1		1.7	0.2	Mn 0.5
NG9/blue	86.0	4.5	0.9		1.2	4.8	1.0	0.1	0.2	0.8			As 0.5
NG9/gilding	32.5	4.8	0.8		0.7			1.1				60.1	
NG8/purple	70.4	16.8	1.1		1.2	3.5	0.8	1.4	3.3			1.5	
NG8/gilding	33.2	4.8	6.4		1.9	0.9	0.5	0.7	3.3			48.1	

[*] Analysts: Edwards and Colomban (2019)

namely cobalt blue, to effect the green colour as there are no temperature stable common green minerals. For example, malachite is a basic copper (II) carbonate and verdigris a basic copper (II) ethanoate. Green earth is a naturally occurring copper ochre, but this is also not stable at high temperatures. Other blue minerals are also unstable at higher temperatures, such as covellite, a basic hydrated copper (II) sulfate, and atacamite, a basic copper (II) carbonate. Egyptian blue, viridian, is a glassy blue cobalt silicate and this too would be stable at the kiln temperatures.

Purple: no Raman spectrum was detected in NG9, which immediately excludes the natural purple mineral *caput mortuum,* a type of finely divided crystalline haematite, and also the compounding of a red and blue colour mixture from haematite and cobalt blue. However, the SEM/EDAXS technique detected the presence of gold in the purple pigment and this immediately identifies it as a *Purple of Cassius* enamel. Metallic gold is not detectable by the Raman spectroscopic technique since it is a monatomic metal and has no chemical bonds: the intensity of scattering in the low wavenumber region of the spectrum can therefore be attributed to Lamb modes of Au^0 nanopairs. An historical survey of the invention of this pigment, made from reacting a solution of stannous hydroxide with chloroauric acid, and its adoption for ceramics decoration is given in Appendix III.

Specimen NG8 gave identical results as NG9 for the purple pigment and also for the gilding. Carbon is found in all pigments and arises from either the deliberate addition of soot as a darkening agent or more likely perhaps from the use of organic adhesives or gums as carriers for the enamel pigments during the painting which would of course burn off to carbon particles at the glost kiln firing temperatures. Also in NG8, several additional Raman bands (at 1017, 1124, 607, 624, 497 and 415 cm^{-1}) are noted for the purple pigment which are attributable to the presence of anhydrite (anhydrous calcium sulfate, $CaSO_4$) which would have been added as gypsum, $CaSO_4 \cdot 2H_2O$, to effect a lightening of the colour tone—the water of hydration ion the gypsum would burn off at elevated temperatures during the firing process to leave the anhydrous anhydrite, a white mineral.

References

J.A. Anderson, *Derby Porcelain and the Early English Fine Ceramic Industry*, Ph.D. Thesis, University of Leicester, UK, October, 2000

I.M. Bell, R.J.H. Clark, P.J. Gibbs, Raman spectroscopic library of natural and synthetic pigments (pre-1850 AD). Spectrochimica Acta Part A **53**, 2159–2179 (1997)

M. Bimson, The examination of ceramics by XRay powder diffraction. Stud. Conserv. **14**, 83–89 (1969)

L. Burgio, R.J.H. Clark, Library of Fourier transform Raman spectra of pigments, minerals, pigment media and varnishes: A supplement to the existing library of Raman spectra of pigments with visible excitation. Spectrochimica Acta Part A **57**, 1491–1521 (2001)

E.A. Carter, M.L. Wood, D. de Waal, H.G.M. Edwards, Porcelain shards from Portuguese wrecks: Raman spectroscopic analysis of marine archaeological ceramics. Heritage Sci. **5**, 17 (2017). https://doi.org/10.1186/s40494-017-0130-9

A.H. Church, Cantor Lectures on some points of contact between the scientific and artistic aspects of pottery and porcelain, Lecture IV, *J. S. Arts*, 126–129 (1880, January 14). Extended in monograph by Trounce Publishers, London 1881

Sir A.H. Church, *English Porcelain: A Handbook to the China Made in England During the 18th Century as Ilustrated by Specimens Chiefly in the National Collection* (A South Kensington Museum Handbook, Chapman & Hall Ltd., London, 1885 and 1894)

P. Colomban, B. Sagon, X. Faurel, Differentiation of antique ceramics from the Raman spectra of their coloured glazes and painting. J. Raman Spectrosc. **32**, 351–360 (2001)

P. Colomban, V. Milande, H. Lucas, On-site Raman analysis of Medici porcelain. *J Raman Spectrosc.* **35**, 68–72 (2004a)

P. Colomban, I. Robert, C. Roche, G. Sagon, V. Milande, Identification des porcelains tendres du 18eme Siecle par spectrometrie Raman: St. Cloud, Chantilly, Mennecy et Vincennes. *Revue d'Archaeometrie*, **28**, 153–167 (2004b)

P. Colomban, V. Milande, On-Site Raman analysis of the earliest known Meissen porcelain and stoneware. J. Raman Spectrosc. **37**, 606–613 (2006)

P. Colomban, The destructive/non-destructive identification of enameled pottery, glass artifacts and associated pigments—a brief overview. Arts **2**, 77–110 (2013)

P. Craddock, *Scientific Investigation of Copies, Fakes and Forgeries* (Butterworth-Heinemann, Oxford, 2009), pp. 201–210

Sir H.T. de la Beche, T. Reeks, F.W. Rudler, *Catalogue of Specimens in the Museum of Practical Geology, Jermyn Street, London, Illustrative of the Composition and Manufacture of British Pottery and Porcelain from the Occupation of Britain by the Romans to the Present Time,* 3rd ed. (George Eyre and W. Spottiswoode, For the HMSO, London, 1876)

L.W. Dillwyn, *Notes on the Experimental Production of Swansea Porcelain Bodies and Glazes.* Made by Lewis Weston Dillwyn with Samuel Walker at the Swansea China Works Between 1815 and 1817. Presented to the Library of the Victoria &Albert Museum, South Kensington, London by John Campbell in 1920. Reproduced in Eccles & Rackham, *Analysed Specimens of English Porcelain*, 1922, see reference below

M.T. Domenech-Carbo, H.G.M. Edwards, A. Domenech-Carbo, J.M. del Hoyo-Melendez , J. de la Cruz-Canizares, An authentication case study: Antonio palomino *versus* Vicente Guillo paintings in the vaulted ceiling of the Sant Joan del Mercat Church (Valencia, Spain). *J. Raman Spectrosc.* **43**, 1250–1259 (2012)

H. Eccles, B. Rackham, *Analysed Specimens of English Porcelain in the Victoria and Albert Museum* (Victoria and Albert Museum, London, 1922)

H.G.M. Edwards, Historical Pigments, in *Encyclopaedia of Analytical Chemistry*, ed. by R. Meyers, Y. Ozaki (J. Wiley and Sons, Chichester, UK, 2015)

H.G.M. Edwards, *Swansea and Nantgarw Porcelain: A Scientific Reappraisal* (Springer, Dordrecht, The Netherlands, 2017)

H.G.M. Edwards, *Nantgarw and Swansea Porcelain: An Analytical Perspective* (Springer, Dordrecht, The Netherlands, 2018)

H.G.M. Edwards, P. Colomban, B. Bowden, Raman spectroscopic analysis of an English soft paste porcelain plaque—mounted table. J. Raman Spectrosc. **35**, 656–661 (2004)

H.G.M. Edwards, P. Colomban, to be published (2019)

M. Hillis, The development of Welsh porcelain bodies, in *Welsh Ceramics in Context, Part II*, edn. by J. Gray (Royal Institution of South Wales, Swansea, 2005), pp. 170–192

E. Jenkins, William Weston Young, in *The Glamorgan Historian*, ed. Stewart Williams, vol. 5. (D. Brown Publishers, Cowbridge, 1968), pp. 61–101. ISBN: 0-900807-8-1

L. Jewitt, *Ceramic Art in Great Britain* (Virtue & Co., London, 1878)

D. Jones, P.E. Jones, Swansea and Nantgarw Porcelains. *The Antique Collector*. 34–41 (1988)

L.D. Kock, D. de Waal, Raman studies of the underglaze blue pigment on ceramic artefacts of the Ming dynasty of unknown origins. J. Raman Spectrosc. **38**, 1480–1487 (2007)

D.A. Long, *The Raman Effect: A Unified Treatment of the Theory of Raman Scattering by Molecules* (Wiley, Chichester, 2002)

E. Morton Nance, *The Pottery and Porcelain of Swansea and Nantgarw* (B.T. Batsford Ltd., London, 1942)

K. Murphy, R. Ramsey, D.A. Higgins, The dismantling of Kiln II, Nantgarw China, Pottery and Pipe Works, Mid-Glamorgan, 1995. Post Mediaeval Archaeology **31**(1), 231–247 (1997)

J. Nickell, *Real or Fake; Studies in Authentication* (University of Kentucky Press, USA, 2009). ISBN-13: 978-081-3125343

M.V. Orna, Chemistry at the Interface of Archaeology and Art, *Chem. Australia*. 470 (1996)

J.V. Owen, R. Barkla, Derby porcelains: Recipe changes, phase transformations and melt fertility. J. Archaeological Sci. **24**, 127–140 (1997)

V. Owen, J.O. Wilstead, R.W. Williams, T.E. Day, A tale of two cities: Compositional characteristics of some Nantgarw and Swansea porcelains and their implications for kiln wastage. J. Archaeological Sci. **25**, 359–375 (1998)

J.V. Owen, M.L. Morrison, Sagged phosphatic Nantgarw porcelain (ca. 1813–1820): Casualty of overfiring or a fertile paste?. Geoarchaeology **14**, 313–332 (1999)

P. Rado, *An Introduction to the Technology of Pottery* (Pergamon Press, Oxford, 1969)

W.R.H. Ramsay, E.G. Ramsay, A Classification of Bow Porcelain from First Patent to Closure: c.1743–1774. in *Proceedings of the Royal Society of Victoria* 119(1), pp. 1–68 (2007). ISSN 0035-9211-1-168

P. Ricciardi, P. Colomban, B. Fabbri, V. Milande, Towards the establishment of a Raman database of early European porcelains. *e-Preservation Sc.* **6**, 22–26 (2009)

P.D.S. St. Pierre, Constitution of bone china I: High temperature phase equilibrium studies in the system, calcium triphosphate—alumina–silica. *J. Am. Chem. Soc.* **37**, 243–258 (1954)

P.D.S. St. Pierre, Constitution of bone china II: Reactions in bone china bodies. *J. Am. Chem. Soc.* **38**, 217–222 (1955)

P.D.S. St. Pierre, Constitution of bone china III: High temperature plasticity studies in a synthetic calcium phosphate-anorthite-silica system. *J. Am. Chem. Soc.* **39**, 147–150 (1956)

J. Taylor, *The Complete Practical Potter* (Shelton, Stoke-upon-Trent, 1847)

M.S. Tite, M. Bimson, A technological study of English porcelain . Archaeometry **33**, 3–27 (1991)

W. Turner, *The Ceramics of Swansea and Nantgarw* (Bemrose & Sons, Old Bailey, London, 1897)

M.E. Wieseman, *A Closer Look: Deceptions and Discoveries* (The National Gallery Company, London, 2010). ISBN 978-185-7094862

I.J. Williams, *The Nantgarw Pottery and its Products: An Examination of the Site*, (The National Museum of Wales and the Press Board of the University of Wales, Cardiff, 1932)

W.W. Young, *The Guide to the Scenery and Beauties of Glyn Neath*, John Wright & Co., Bristol (sold by Longman, Rees, Orme, Brown & Green, Paternoster Row, London, 1835)

W. W. Young, *The Diaries of William Weston Young, 1776–1847(1802–1843)*, 30 volumes, West Glamorgan Archive Service, Swansea, SA1 3SN. https://arcgiveshub.jisc.sc.uk/data/gb216-d/dxch/ddxch/i/hub

Chapter 11
Epilogue

Abstract A review of the discoveries made by William Weston Young and a summary of his influence upon the two extreme ceramic examples of fine porcelain and silica refractory bricks.

Keywords William Weston Young · Nantgarw porcelain · Dinas silica refractory bricks · Hard paste porcelain · Dry mix porcelain powder

In recounting the history of William Weston Young and his two ceramic extremes several important milestones have become apparent: his role in the creation of the Nantgarw China Works and his determination for it to continue in adverse circumstances after the premature departure of William Billingsley and Samuel Walker for Coalport is now firmly established. Hitherto, in previous accounts although much of the credit for the achievement and perfection of Nantgarw china, accorded the description of "*the world's finest china*", is fairly attributed to the efforts and drive of Billingsley and Walker the support of Young in their endeavours has often been minimised or even ignored. Indeed, it is clear that without the financial backing and supportive personal introductions to local sponsors effected by William Weston Young, William Billingsley and Samuel Walker would not have acquired the wherewithal to embark upon porcelain manufacture at Nantgarw upon leaving Messrs. Flight, Barr and Barr at Worcester in 1813. This was particularly evident for the re-commencement of porcelain manufacture at Nantgarw in the so-called phase II operation, which commenced in 1817 after their failure to secure government funding of the initial phase in September 1814. It was purely through Young's personal financial commitment and his exhortation of support from local landowners and businessmen that the Nantgarw China Works was able to re-start the production of porcelain, which resulted in a phenomenal demand for its china from a discerning London clientele which surpassed its ability to provide it in sufficient quantity to establish a viable commercial base due to excessively high kiln losses, which approached 90%. This unacceptable commercial enterprise resulted in its failure after only two years' full production and the departure of William Billingsley and Samuel Walker to the Coalport China Works of John Rose in 1820. This did not deter William Weston Young in his ambition to keep the Nantgarw China Works open and he engaged the accomplished ceramics enameller Thomas Pardoe to assist him in

© Springer Nature Switzerland AG 2019
H. G. M. Edwards, *Porcelain to Silica Bricks*,
https://doi.org/10.1007/978-3-030-10573-0_11

this. Young accepted the responsibilities of running the operation and the recurrent debts of the Nantgarw China Works and succeeded in defraying the advances of the creditors by preparing with Pardoe the remaining stock of biscuit and glazed china ware left at Nantgarw for sale locally: this Nantgarw china is particularly highly prized by collectors today, especially the finer pieces decorated by Pardoe with his superb floral arrangements, birds and insects (John et al. *Nantgarw Porcelain Album* 1975). Examples include the *Wyndham Lewis* and *Spence-Thomas* services, which are illustrated here in Figs. 4.1 and 4.3, respectively. Although the taking over of the remnant stock at the Nantgarw China Works for decoration by Young and Pardoe has been documented widely hitherto, what has not been so well appreciated and highlighted historically is the effort expended by Young to re-create Nantgarw porcelain production and the rekindling of the furnaces there in 1820—and, moreover, how close he came to its realisation, being defeated only through a lack of appreciation of the ingredients and composition of the porcelain body that was kept secret for so long by Billingsley and Walker.

A search of the historical literature reveals that William Weston Young developed a lifelong interest in ceramics decoration from his earliest years, which can be traced back to his association with Lewis Weston Dillwyn at the Cambrian Pottery in Swansea in 1803, immediately following his disastrous venture into the general goods merchant business which eventually bankrupted him. It was here too that he met and nurtured his friendship with Thomas Pardoe; Young was decorating Swansea earthenware at the Cambrian Pottery and after he left in 1806, he also enamelled and glazed pieces of Swansea earthenware using his own muffle furnace at Newton Nottage, whilst pursuing his other interests in salvaging wrecks and land surveying. His ceramics decorating business continued for the next decade until he then met up with William Billingsley and Samuel Walker at Nantgarw and the triumvirate submitted their *Memorial* document in September 1814 to the government in the crucial but unsuccessful request for funding to launch the Nantgarw China Works enterprise. It is an interesting diversion to explore the different attitudes prevailing between Billingsley and Walker on the one hand and Young on the other from their equable partnership in 1814 to the eventual departure of Billingsley and Walker from Nantgarw for Coalport in 1820, clearing their own sheet and leaving Young with all the responsibilities and debts accrued by the Nantgarw China Works and yet patently not assisting him in his continuation of porcelain production there by giving him the formulaic recipe either for the synthesis of the porcelain itself or of its glaze. Young, assisted by Pardoe, managed to create a new glaze (the so-called Nantgarw No. 2 glaze) for the Billingsley/Walker biscuit ware remnants left at Nantgarw but he lacked sufficient information to make the soft paste porcelain and instead tried to manufacture an inferior hard paste body. It is a matter for conjecture as to the reason that Billingsley and Walker never divulged the secret Nantgarw porcelain recipe to their partner, Young, who nevertheless took on their debts and freed them up at Coalport—which implied that they certainly did not part company on bad terms with each other. Perhaps, Billingsley and Walker were still cognisant of the rather threatening letter from Flight, Barr and Barr at Worcester sent to Walker and Dillwyn at Swansea just six years earlier, cautioning against the release of sensitive information about the composition of their porcelain body to a "third person"—even though that person was at that time a full partner with them in porcelain manufacture at Nantgarw! In retrospect, it does seem a rather poor reward for Young's kindness and considerable financial aid over the preceding three years or so, including £1100 of his own money, without which the Nantgarw China Works would not have been opened up for its phase II operations in 1817; an alternative view is that Billingsley and Walker were saving Young from the impending wrath of Messrs. Flight, Barr and Barr, who would have brought down a legally enforced embargo upon Young's fledgling operation at Nantgarw and effectively closed him down forever with a swingeing financial penalty and presumably legal costs that he could not entertain to meet. Nevertheless, after the untimely death of Thomas Pardoe in 1823, Young's business enterprise at Nantgarw was in serious financial trouble and he became bankrupt yet again for the second time—whether or not the revelation of the Billingsley and Walker formulaic recipe for Nantgarw porcelain would have saved him is doubtful since all three could not make it a commercially viable exercise in the first place because of the appallingly high kiln losses suffered, and this would surely have been the situation under Young's tenure after 1820 anyway. The persistence of William Weston Young in attempting to continue with the porcelain production at the Nantgarw China Works from 1820 after the departure of William Billingsley and Samuel Walker is a tribute to his determination and valiant effort to turn around the business fortune but unfortunately, like some of his other business ventures, it resulted in a commercial disaster and his inability to recover from the outstanding debts and it folded in 1823, making Young a bankrupt for the second time. The question remains as to how much porcelain was made by Young at Nantgarw and was this ever a successful commercial venture? There have been suggestions by previous historians that Young and Pardoe must have bought in a supply of porcelain form other factories to supplement the remnant Nantgarw porcelain left in stock by Billingsley and Walker—much of this must surely have been of "seconds" quality as John Mortlock had arranged with Billingsley to take all the perfect china that he could supply for the London market where it could be decorated in the ateliers of Robbins and Randall and John Sims and then sold on at a premium. The decoration of "second", and even "third" and "fourth" grade porcelains was a common practice at this time in other factories such as Derby (Anderson 2000) as it was a means to recoup revenue form only slightly deformed china, which may also have had some firing cracks or blemishes which could be masked by local enamelling and glazing procedures. An example of such a case for Nantgarw is provided in Fig. 5.1a, b and c, where a dessert plate which has some surface blemishes and some slight warping has been decorated locally during the Billingsley and Walker era, as evidenced by the No. 1 glaze and gilding (see Appendix D). Others have also hinted that the quantity of china left in stock, in biscuit or in glazed form, from the Billingsley and Walker era would not have been sufficient to occupy Pardoe's enamelling talents fully and, therefore, there would have been much to recommend the external purchase and importation of porcelain from elsewhere. This statement cannot be verified independently without the existence of written records so it must be just logged as a possibility without any evidential material in support although it is recorded that some 15 years hitherto Pardoe and Young had collaborated very successfully in the decoration of earthenware at the Cambrian pottery and it is known that Pardoe had a very successful china enamelling business established in Bristol right up to his recruitment by Young to participate in the Nantgarw enterprise in 1821.

Young's experiments in the recreation of Nantgarw china began in early 1819, interestingly undertaken chronologically before Billingsley and Walker had left the factory, and these are recorded in detail in his diaries; it could be surmised that at this stage Young had learned of his partners' wishes to leave Nantgarw and he was therefore undertaking some prior ceramic work to facilitate his ability to continue when they eventually departed. It is clear that Young was poised to put in a bid for the Nantgarw site through the landlord Edward Edmunds and to retain the stocks, manufacturing hardware and material there—this actually happened in September 1820 (a copy of this contract between Edmunds and Young still exists), not long after Billingsley and Walker had left Nantgarw, so we can surmise that by that stage Young had objectively assessed the results of his experimental porcelain body work and judged that he could reasonably plan to continue the operation at Nantgarw. We can now hypothesise if Billingsley and Walker actually knew of Young's intentions at this stage or not and if this influenced their decision to keep their formulation secret from him—what is clear is that Young was able and willing to take on the Nantgarw debts and site management, and even hire Pardoe to effectively move forward and maintain the Nantgarw China Works in operation almost immediately. It is doubly interesting that Young had used the Nantgarw kilns for the firing of his "*ersatz Nantgarw*" porcelain recipe over a year prior to his taking over the China Works and it would have been illuminating if he had in fact received prior knowledge of the true Billingsley/Walker porcelain body formulation, which included the significant bone ash ingredient that he obviously lacked in his own preparation. Even though Young's efforts to reproduce Nantgarw china failed during his tenure at the Nantgarw China Works, after leaving in 1823 he still persisted in attempting to market his version of Nantgarw china: records and diary entries show that he engaged with colleagues at the Glamorgan Pottery in Swansea and used their ceramic kilns to fire his porcelain mixtures in efforts to improve the body composition. This resulted in a change of tack somewhat as Young then marketed his "*ready-mix*" Nantgarw porcelain powder which required the addition of water before firing and glazing. After some initial interest and purchases made by Staffordshire ceramics factories, the Nantgarw *ready-mix* composite was deemed to be inferior in translucency and overall quality to the Billingsley/Walker porcelain body, which is what the potential market required despite it being judged a "tolerably good porcelain", so the high expectations of the *ready-mix* purchasers were not achieved. This situation persisted through the 1830s and even into the 1840s as Young never gave up trying to sell his *ready-mix* porcelain.

In contrast with the compound difficulties that William Weston Young experienced in taking over the Nantgarw China Works and the decoration of the remnant stock, in trying to re-create Nantgarw porcelain and in marketing his *ready-mix* porcelain powder under the guise of Nantgarw porcelain, the comparative ease in which he launched his invention and manufacture of the *Dinas* silica refractory brick and its international success is quite remarkable. Within two years of carrying out his initial experiments on the formulation and firing of silica bricks at the Nantgarw China Works he moved into their production at his Pont Walby site in the Vale of Neath with immediate national acceptance of an exclusive and fine product which spread internationally and gave him the market edge with little or no competition, a situation that was so very different form that which he found with Nantgarw porcelain. Although he could not officially be in command of the operation at the *Dinas* Silica Brick works, he was able to rent the site from the Marquess of Bute in 1822 and started the manufacturing process in 1823, just as the Nantgarw China Works was being wound down through the final auction sales. In its genre, the *Dinas* refractory silica brick is still recognised as the world's finest of its type—a position identical with that occupied by Nantgarw porcelain in its own genre! His rewards from the Dinas Silica Brick Works soon enabled Young to discharge his bankruptcy obligations after his lack of success in the Nantgarw operation, and he accomplished this in 1828. Thereafter, he handed over the reins to his eponymous nephew at the Dinas brick works and removed to Bristol to further his other interests—the major one being a further series of attempts to sell his Nantgarw porcelain formulaic version, seemingly with varied success. The unanswered question, of course, is whether or not William Weston Young ever succeeded in making a quantity of "Nantgarw" porcelain: seemingly he did, because documentation exists that a potential purchaser of his powdered formulation asked him to provide several finished and decorated examples of his wares—which he was able to do through a re-purchase of porcelain sold to locals in Neath some years earlier, where he was living at the time. This is an intriguing situation because it is true that we would not recognise these items as Nantgarw porcelain today, being hard paste and probably reconciled and assigned to French or Chinese copies made later in the 19th Century. It is even recorded that Young impressed the mark NANTGARW into his flatwares, but that this was smaller than the Billingsley/Walker original impressed mark shown in Fig. 1.4 and was additionally contained within a rectangular box-like cartouche (Fig. 10.7).

From a socio-historical point of view it is a fact that all four of the major characters discussed here, namely William Weston Young, William Billingsley, Samuel Walker and Thomas Pardoe, died in rather poor circumstances—the first two having no issue. William Weston Young died childless but his wider family descendants have provided some records that have assisted historians in an appreciation of his work: Elis Jenkins (*Glamorgan Historian* 1968) refers to a Colonel Young of Preswylfa, Neath, and a Miss Young who had a small collection of her ancestors' ceramic works and it would be revealing if one or more of these could be assigned to William Weston Young's attempts at porcelain manufacture at Nantgarw!

11.1 Nantgarw Porcelain: An Invariant Composition?

It is very relevant here to consider the statement made by previous historians of the factory that the Nantgarw China Works manufactured porcelain of an *invariant composition* during its very short lifetime. The originator of this statement seems to be Herbert Eccles, who examined the collection of Nantgarw and Swansea porcelains which was accumulated for the Centenary Exhibition at the Glynn Vivian Art Gallery in Swansea in 1914 to celebrate the first production of porcelain at the Swansea China Works by Lewis Weston Dillwyn in 1814. Eccles was able to declare from his studies of the Nantgarw pieces on show there that, unlike the Swansea china, all Nantgarw china had an identical composition. This remark was made by visual examination only and did not involve any form of scientific, chemical analysis, probably because all the porcelain was in the form of perfect pieces and analytical chemistry would require the destruction of s significant part of the specimens. This was undertaken some eight years later in 1922 when Eccles and Rackham (*Analysed Specimens of English Porcelain* 1922) used wet chemical acid digestion to estimate the chemical composition of three pieces of Nantgarw porcelain from the Victoria and Albert Museum's *Lady Charlotte Schreiber Collection*—these chemical analyses vindicated Eccles' earlier proposition that the Nantgarw porcelain composition was *invariant*. Several important milestones have appeared during this current investigation, the implications of which relating to Nantgarw porcelain composition should now be expanded on and discussed.

The Phase I Nantgarw China Works operation, which commenced in 1814, wherein Billingsley, Walker and Young first created Nantgarw china and submitted examples of finished and decorated specimens for evaluation in support of a submission for government funding as detailed in their *Memorial* of September 5th, 1814: it is not known how many pieces of Nantgarw china were made in this phase and it is estimated that perhaps five or so pieces and no more would have been submitted along with the *Memorial* document. What is unknown is the number of pieces manufactured altogether in this opening phase: typical kiln loadings for this period were some 25 dozen pieces in full production, so it is reasonable to suggest that in view of the high kiln firing wastage experienced at Nantgarw that a batch of at least several dozen pieces would have been made and fired, and perhaps even more when one considers that it would have taken some trial runs and several batches before the temperature and kiln firing times were correctly determined. There is no record, however, of any porcelain sales being confirmed in nay documentation during this phase I period at Nantgarw, but a few dozen perfect pieces may have survived from which a small sample would have been selected for the *Memorial* submission. It is accepted that the recognition of these first pieces, although presumably carrying an impressed Nantgarw mark, now would be a difficult undertaking and perhaps problematic, especially if the porcelain body was modified further by Billingsley and Walker after their unsuccessful submission was realised. This would be the case if a later inspection would have denied their origin and assigned them to a hybrid type of porcelain and therefore to foreign copies, such as the later hard porcelain fakes perpetuated by firms such as Samson of Paris in the mid to late 1800s (Edwards, *Swansea and Nantgarw Porcelain: A Scientific Reappraisal* 2017).

Was Nantgarw porcelain made by Lewis Weston Dillwyn at the fledgling Swansea China Works during the early part of 1815, just after he recruited Billingsley and Walker to manufacture porcelain for him at Swansea? Some serious consideration has been afforded to this concept by earlier authors which has been reinforced by the suggestion that as Dillwyn was starting up his porcelain factory from scratch he would have needed a formulation for his Swansea body paste and it has even been suggested that he could have used the china moulds brought from Nantgarw by Billingsley and Walker for this purpose. Morton Nance (*The Pottery and Porcelain of Swansea and Nantgarw* 1942) explores this idea and considers it as a reasonable explanation for the presence of certain unusual porcelain plates made with the characteristic Nantgarw mouldings, and marked with the correct impressed Nantgarw mark, but decorated by Swansea artists such as William Pollard and Henry Morris, who never were employed at the Nantgarw China Works! An alternative explanation is rather difficult to propose to counter this, except for a suggestion that perhaps these items were purchased privately at one of the Nantgarw China Works auction sales in 1822 and 1823 for decoration at home by Pollard and others: a critique of this proposal is that the artists would have been required to purchase the Nantgarw porcelain in biscuit and undecorated form, whereas it is known that Young and Pardoe were intent upon decorating their remnant stock before sale to achieve the best prices. William Billingsley is recorded as saying in a letter to John Coke, his Principal at the Pinxton China Works, that it was not commercially viable to sell porcelain in the white and undecorated (Anderson 2000), which brought in a pittance in comparison to the decorated pieces. Yet, in contrast, it is well known that factories did sell a small amount of porcelain and earthenware ceramics in the white to private decorators and enamellers, who then completed the work and glazed the specimens in their own muffle furnaces: William Billingsley did this to great effect at Mansfield and Brampton-in-Torksey, and William Weston Young at Newton Nottage.

The Phase II operation at Nantgarw, which occupied the period from 1817 to 1819/20, is the normally accepted production period associated with the Nantgarw China Works, wherein the porcelain produced was accorded the title of "the best porcelain in the world". The assumption has always been made, from the initial observations of Herbert Eccles in 1914, that the body was *invariant,* that is, no changes were made to its composition during this production period and most of the Nantgarw porcelain that we see today emanates from this narrow production period and the intense activity of Billingsley, Walker and Young. Certainly, most of the shards that have been analysed to date from the Nantgarw China Works site belong to this constant body composition period, However, in a very few instances it seems that a departure from this body composition has taken place and this will be elaborated upon later. More recently, the non-destructive spectroscopic analyses of perfect finished Nantgarw porcelains from this period also indicate that the body composition is unchanging. It must be remembered that realistically the sampling of the shards and perfect porcelain items represents just a small percentage of the existing porcelain: Dr John estimated that he had personally viewed most of the Nantgarw porcelain in existence when compiling his book (*Nantgarw Porcelain* 1948) and that this amounted to approximately 5000 pieces. In comparison, only some four pieces of perfect china and some thirty shards of Nantgarw Porcelain have ever been analysed chemically less than one per cent of all the porcelain identified by Dr John.

   The Post-Phase II Nantgarw China Works period, occupying from 1820 to 1823, is the period when William Weston Young and Thomas Pardoe attempted to revive the flagging fortunes of the Nantgarw China Works by decorating the remnant stock left in store by Billingsley, Walker and young from the phase II operation. It is the general consensus of opinion form previous historical accounts that no porcelain was ever made again at Nantgarw during this period and that Young and Pardoe devoted their efforts to enamelling the stock items left from 1817 to 1819 in biscuit or glazed and selling these privately and locally and through the auction sales held during 1822 and 1823. Generally, experts can easily identify Nantgarw porcelain from this period from possible slight defects which would have rendered it unfit for acceptance by Mortlocks for the London market and by the use of the Nantgarw No. 2 (Young/Pardoe) glaze, which was creamier and thinner than that used by Billingsley and Walker. Further clues can be found in the identification of Pardoe's characteristic hand in the enamelling and also in the scarcity of the applied gilding—this was considered one of the most expensive parts of the decorating process and there are cases recounted of services commissioned from factories economising on the gilding at the request of the customer to save on costs. Classic examples are the *Duke of Devonshire* service commissioned from the Derby China Works (Edwards, *Swansea and Nantgarw Porcelain*: *A Scientific Reappraisal* 2017; Twitchett, *Derby Porcelain*: *1748–1848*: *An Illustrated Guide* 2002) and locally decorated Nantgarw services where the normal edge gilding has been replaced with coloured enamels such as blue, green and chocolate. William Weston Young visited John Mortlock in London in 1821, who was the sole Nantgarw China Works supplier in the capital under the tenure of Billingsley and Walker, and tried to persuade him to take some Nantgarw "seconds" decorated by Pardoe but, disappointingly for Young, Mortlock refused to do so: his clientele had always demanded the highest quality china and were prepared to pay a premium for this—clearly, china which exhibited firing cracks, blemishes and warping would not be acceptable and this was the reason for Billingsley and Walker not sending these pieces to London in the first place. In the 1823 auction sale the final disposal of Nantgarw china was accomplished, Thomas Pardoe had meanwhile died in the July and Young then departed the Nantgarw China Works site, never to return.

   During this Post—Phase II Nantgarw period described above, a particular interest centres upon the years 1821 and 1822, when Young records his experiments in his diaries on trial experiments in formulating a new Nantgarw body in emulation of that of Billingsley and Walker—unfortunately, he had no knowledge of the necessity of inclusion of the bone ash component which was so vital in the original formulation and his formulaic recipe is better described as a hard paste porcelain, really rather close to the Chinese type. Again, Young could not persuade potential clients to purchase this in quantity but it appears that some sales were made locally—which affirms that, firstly, porcelain was made at Nantgarw after Billingsley and Walker had left for Coalport and, secondly, examples survived in clients' possession for a while afterwards as Young recorded that he was able to re-purchase some for the exhibition of his finished china to potential customers some years later. In this context it is interesting to note that some earlier accounts (Morton Nance, *The Pottery and*

Porcelain of Swansea and Nantgarw 1942; Turner, *The Ceramics of Swansea and Nantgarw* 1897) do actually refer to the possibility that Young tried to continue with porcelain production at Nantgarw after the departure of Billingsley and Walker in 1819/20. Morton Nance states that he used an impressed mark, rather larger than that associated with the Nantgarw China Works during the Billingsley/Walker tenure, which was enclosed in an impressed rectangular box cartouche: the author has devised such a mark which is shown in Fig. 10.7. Recently, it appears that a "Nantgarw" plate offered at auction was typically of the shape and moulding and high quality decoration expected from the factory but suffered from two problems—firstly, it comprised "an English hard paste porcelain" and bore a rather strange impressed Nantgarw mark, without a hyphen and missing the characteristic C.W. symbols, in a box cartouche, similar to that shown in Fig. 10.7. The plate was judged by expert opinion; quite correctly it seemed at that time, to be a fake Nantgarw item and was withdrawn from the auction. However, it is quite feasible that this plate could be a very rare survivor of William Weston Young's efforts at porcelain manufacture at the Nantgarw site in the post-Billingsley and Walker period of tenure. Forensically, we are now also faced with an interesting impasse: namely, if William Weston Young did in fact create a hard paste porcelain at the Nantgarw China Works, using the old Nantgarw moulds and decorating it beautifully in the accepted style (he was after all an accomplished ceramic artist in his own right) after firing the biscuit and glazing it in the Nantgarw kilns was he not able to correctly describe and claim it as "Nantgarw porcelain"? The problem is, of course, that clients wished to purchase Billingsley and Walker's Nantgarw porcelain, and their remnant stocks decorated by Pardoe and Young—and this was very different form Young's creation. As an afterthought, other factories during their operational lifetimes occasionally made significant alterations to their body formulations—a good example being that if the Swansea China Works which manufactured three identifiably different porcelain bodies between 1815 and 1819, namely, the *glassy, duck-egg* and *trident* types, which are still all acceptable as *Swansea porcelain*! So, Young's porcelain, if and when examples surface once more, should not perhaps be condemned as *fake Nantgarw,* but rather be regarded as rare examples of true Nantgarw porcelains (and hence very worthy of inclusion in a museum collection), which may still be catalogued as not as desirable for collectors as the Billingsley and Walker type, but like Swansea *duck-egg, glassy* and *trident* bodies, perfectly acceptable as different examples of the genre.

The period 1854–1858 at Nantgarw represents the final occasion when porcelain was made at the China Works site: this being a hard paste porcelain manufactures by William Henry Pardoe, son of Thomas Pardoe. The final sale of this "Nantgarw china" along with the moulds, manufacturing equipment and materials was advertised by Pardoe in 1858—so several years' production must have been undertaken. To the authors knowledge none of this porcelain has been identified to date and attributed to the William Henry Pardoe era and, additionally, analyses of the shards from the corresponding area of the waste pit identified with Pardoe's earthenwares (Williams 1932) has never been undertaken. This would be necessary to attempt to define the hard paste porcelain composition of Pardoe's porcelain from the 1850s.

References

J.A. Anderson, *Derby Porcelain and the Early English Fine Ceramic Industry*, Ph.D. Thesis, University of Leicester, UK, October, 2000

H. Eccles, B. Rackham, *Analysed Specimens of English Porcelain in the Victoria and Albert Museum* (Victoria and Albert Museum, London, 1922)

H.G.M. Edwards, *Swansea and Nantgarw Porcelain: A Scientific Reappraisal* (Springer, Dordrecht, The Netherlands, 2017)

E. Jenkins, William Weston Young. *The Glamorgan Historian*, ed. by Stewart Williams, vol. 5 (D. Brown Publishers, Cowbridge, 1968), pp. 61–101. ISBN: 0-900807-8-1

W.D. John, *Nantgarw Porcelain* (Ceramic Book Co., Newport, 1948)

W.D. John, G.J. Coombes, K. Coombes, *The Nantgarw Porcelain Album* (Ceramic Book Co., Newport, 1975)

E. Morton Nance, *The Pottery and Porcelain of Swansea and Nantgarw* (B.T. Batsford Ltd., London, 1942)

W. Turner, *The Ceramics of Swansea and Nantgarw* (Bemrose & Sons, Old Bailey, London, 1897)

J. Twitchett, *Derby Porcelain, 1748–1848: An Illustrated Guide* (Antique Collectors Club, Woodbridge, Suffolk, 2002)

I.J. Williams, *The Nantgarw Pottery and its Products: An Examination of the Site*, (The National Museum of Wales and the Press Board of the University of Wales, Cardiff, 1932)

Appendix A

A.1 Transcript of Letter from Flight, Barr & Barr, Worcester, to Samuel Walker, dated November 12th, 1814, with Notes and Commentary

Mr. S. W.   Worcester. Nov: 12th, 1814.

Sir

We were a good deal surprised after the kind and liberal treatment yourself and Mr. Billingsley received from us, that you both so suddenly left our works—how far this conduct was consistent we leave you to consider—in addition however to this breach of confidence, we are now told that you are about forming some sort of connection with a Person with the name of Young, and also with Messrs. Dillwyn and Bevington Potters of Swansea, and that you are to make for them a composition the principles of which are similar to the one for which we paid you a high premium, besides being at great expenses in your wages etc. during the time of your acquiring this knowledge by experiments made at our works.

You well know that you engaged that the secret should be entirely confined to ourselves, or ultimately to you and Mr. Billingsley if ever you should venture to make the Article yourselves, *but not to the profit of any Partners, or yourselves if engaged with other Partners*—for is this were the case the advantage to be derived from the secret, for which we gave you so large a price would be destroyed.

You and Mr. Billingsley are jointly and severally bound to us in a penalty of one thousand Pounds to forbear from communicating the secret to any person or persons whomsoever, and is we find upon further enquiry that you really mean to adopt a line of conduct so dishonourable as that before alluded to—we now inform you of our firm resolution of instantly giving our attorney Instructions to commence an

© Springer Nature Switzerland AG 2019
H. G. M. Edwards, *Porcelain to Silica Bricks*,
https://doi.org/10.1007/978-3-030-10573-0

Action against you for the amount of the Penalty of one Thousand Pounds named in
the Bond given to us the 17th day of November 1812.
We are Sir
Yrs etc.
Flight, Barr & Barr

PS. We shall wait the return of the Post for your answer before we address a letter
on this subject to Messrs. Dillwyn and Bevington.

A.2 The Bond

Know all Men by these presents that we … are held and firmly bound to… in eth
Sum of one thousand pounds of good and lawful money of Great Britain to be paid
to the said … or their Attorney, Executors , administrators or assigns for eth true
payment whereof we bind ourselves jointly and severally and our respective heirs
executors and administrators for and in the whole firmly by these presents sealed
with our seals dated this 17th day of November in the 53rd year of the reign of
George 3 and of our Lord 1812, whereas in consideration of the sum of £200 of
Lawful British Money now paid the receipt where of is hereby acknowledged and
also divers sums of money paid in the course of the last three years by way of wages
while Experiments were making they the above bounden … have imparted and
disclosed to the said … the knowledge of a certain secret relating to a new method
of composing Porcelain the principles of which are specified and set forth on the
back hereof and have agreed not to reveal the same to any person or persons
whomsoever or to make use thereof to their own advantage in any manner—Now
the considerations of this obligation is such that the above bounden … Shall and do
form time to time and at all times forever hereafter forbear from communicating and
imparting the secret above mentioned to all and every person or persons whomsoever and also shall and do from time to time and at all times refrain from making
use or availing themselves of their knowledge of such a secret in order to procure
any further emolument or advantage whatsoever unless they or either of their Heirs
shall at any time hereafter engage in the Trade or Business of a China Manufacturer
in which case they reserve to themselves the Liberty of making use of the said
secret in carrying on such Business then this obligation shall be void but otherwise
to be and remain in full force and effect.
    Sealed and Delivered being first duly stamped in the presence of …

A.3 Notes on the Flight, Barr & Barr Letter to Samuel Walker

This letter is important forensically for several reasons as it dispels some myths associated with the relationship between Billingsley and Walker and firstly, Barr, Flight and Barr and then Flight, Barr & Barr at Worcester and, secondly, the employment of Billingsley and Walker by Dillwyn at Swansea after the failure of their phase I operations at Nantgarw. The first point we can make concerns the chronology of events surrounding their departure and subsequent engagement at Swansea: the original Bond drawn up at Worcester between Billingsley, Walker and the then proprietor, Martin Barr, was dated the 17th November, 1812, and in this the sum paid upon their severance was specifically mentioned at £200. The implication is clear in that Billingsley and Walker for the previous three years had been engaged by Barr to develop a new porcelain and process. Several authors have stated that Billingsley and Walker were already based at the Royal Worcester China Works in 1808, after they had moved there from Brampton-in-Torksey, where Sarah and Lavinia Billingsley are recorded as being employed as burnishers in the factory enamelling workshop; Samuel Walker married Sarah Billingsley there in 1812. Hence, it is intriguing to note that Billingsley and Walker … must have been employed at Worcester on other duties prior to their engagement by Martin Barr to develop a new porcelain body.

The chronology of the Letter is interesting as it was written on November 12th, 1814, almost exactly two years after the original Bond affirmed between Martin Barr of Barr, Flight and Barr, William Billingsley and Samuel Walker. Billingsley and Walker are believed to have departed the Worcester China Works in 1813 after the death of Martin Barr and because the ascendancy of George Flight leading the new Flight, Barr & Barr consortium dictated that they would not favour financially embarking upon the development of a new porcelain for Worcester. It is interesting that Billingsley and Walker's setting up at Nantgarw China Works with William Weston Young did not seem to cause Flight, Barr & Barr any concern and it was not until Billingsley and Walker were taken on by Lewis Dillwyn after their failure to draw down government support in early September of 1814 generated their letter to Samuel Walker that is reproduced here. So, we can surmise that Flight, Barr & Barr saw the danger in the link-up with Dillwyn and attempted to frighten off Walker in this proposed enterprise.

The threat at the end of Flight, Barr & Barr's letter to inform Dillwyn of a potential punitive legal action against the Swansea China Works is a real one and we learn from elsewhere that they did warn off Dillwyn and attempted to discourage him from engaging Billingsley and Walker. Dillwyn did not comply with their terms but it was not long before, just several months later, that he started to experiment himself in the production of a new Swansea body: his work books from August 1815 remark upon the achievement of the esteemed, splendid duck-egg porcelain for which he is recognised. it is believed that prior to this, the Swansea body was a glassy porcelain, often rather heavily potted and with a thick glaze, even so it attracted much of Billingsley's early artistic work at Swansea, until the duck-egg porcelain assumed its output majority.

An intriguing point is that the Letter from Flight, Barr & Barr is addressed to Walker and not to Billingsley, although he is mentioned within it along with Young. One could infer from this that Walker was perceived as the practical brains behind the new porcelain body, and not Billingsley, who was nevertheless appreciated for his artistic and ceramics decorating skills.

The Letter also contains a phrase, delineated in italicised script, which does not feature at all in the original Bond and must therefore be described as a Flight invention to try and sequester the activities of Billingsley and Walker even more by forbidding them to associate with any Partners for the production of their new porcelain—thereby directly involving Dillwyn in the legal argument! That nothing came of the case is evidence that Flight, Barr & Barr could not foresee their achievement of the penalty clause and fine of £1000—a very significant and considerable sum at that time, possibly approaching £1M today. A theory that has been proposed for some time is that Flight, Barr & Barr knew well of Billingsley and Walker's initiation of Nantgarw China Works but that they only became worried when Dillwyn came on the scene as they realised that he had the wherewithal financially to promote porcelain manufacture which they did not have at Nantgarw. Hence the stress in their Letter first to Walker and subsequently to Dillwyn that the collaboration with Partners in the endeavour would lead to legal prosecution, which never featured in Martin Barr's original Bond notice. Whatever the true situation was, it seems that the tone of the Letter from Flight, Barr & Barr to Walker was heavily threatening rather than politely conciliatory and this indicated the presence still at Worcester of some upset and bad blood ... the fact that Billingsley is referred to as a "rogue" is indicative of this as is the detail that the Letter was written to Walker and not Billingsley—maybe appealing to his better judgement and perceived social conscience?

A particular bone of contention arises around some potential industrial espionage that enabled Flight, Barr & Barr to learn of Dillwyn's intention to engage Billingsley and Walker at Swansea so soon after the collapse of their unsuccessful application for government support in September, 1814 (*The Memorial*). We do not know how Flight, Barr & Barr learned of Dillwyn's intention to employ Billingsley and Walker at Swansea, or even how they traced them there—unless there was a possible leak of information at government level which tipped them off at the Royal Worcester China Works that a potentially dangerous rival and competitor was emerging on the scene at Nantgarw—although the project application was unfunded the proximity of such expertise to Dillwyn's avowed intention of manufacturing porcelain at Swansea would not have been lost on them? In fact, the phenomenal success of the Swansea duck-egg soft paste porcelain body was a real threat to Worcester's hard paste and soaprock body, which was certainly more robust but generally not deemed to be as attractive.

The question as to whether or not Lewis Dillwyn did manufacture a Nantgarw recipe body at Swansea cannot be definitively answered: several authors have assumed that Billingsley and Walker would have informed him of their formulation as part of their engagement deal at the Swansea China Works, but we have seen how jealously this formulation was guarded—and even their major, and indeed only sponsor for the Nantgarw operation in 1814 was William Weston Young and even he was not aware of the formulation recipe! Another suggestion has been that Walker, who was very closely involved with Dillwyn's chemical fine tuning of the body compositional materials, could have "steered" him down a parallel path without actually divulging the details of eth Nantgarw formulation. Most ceramic historians believe that the earliest Swansea paste formulation was glassy, which included a proportion of glass frit or cullet to increase the transparency of the melt. This was a common additive in the 18th Century earlier porcelains, such as Chelsea and Bow, and did maintain into the 19th Century until the advent of the bone ash and china clay composite was adopted more widely. However, some oddities form an apparently early Swansea production are known which do not meet these criteria—in particular, plates of a Nantgarw moulding which have been decorated by Swansea artists such as Morris and Pollard, who never worked at Nantgarw; it has been proposed that some of the earliest kiln firings at Swansea therefore, comprised Nantgarw paste or even biscuit porcelain remnants, with the impressed Nantgarw mark!

Appendix B

Notes on the Experimental Production of Swansea Porcelain Bodies and Glazes Made by Lewis Weston Dillwyn with Samuel Walker at the Swansea China Works Between 1815 and 1817. Presented to the Library of the V&A Museum by John Campbell in 1920. Taken with modification from Eccles & Rackham, Analysed Specimens of English Porcelain, 1922.

Key to abbreviations in the original text:
V sand; KO flint; LO lime; YX bone; B St. Stephen's clay; E Norden clay; FO composition, china stone, FX pearl ash; EX nitre; GX arsenic; AX lead; MX borax; DX glass; LX smalt; SR soaprock; No. 157 sand frit.

Body Number 1:
Porcelain body
12V sand + 1 FX pearl ash—fine
10 FO china stone + 1 FX pearl ash—coarse
No glass content.

Body Number 2:
Glass frit: 11 V sand + 9 FO china stone + 6 FX pearl ash + 3 MX borax—26 parts taken with 12 AX lead + 1 SR soaprock
The first mention of a glass frit with high lead oxide content.

Body Number 3:
3 V sand + 3 FO china stone + 2 FX pearl ash, fritted with one-tenth of SR soaprock
The above is a "Variation from the Nantgarrw body". (*Author's note: Nantgarw porcelain does not contain any glass frit or lead glass component — is this what Dillwyn means by a variation from Swansea in the Nantgarw body and does this imply that he knew the Nantgarw porcelain formulation or that he had received some lead or inkling from Walker alongside him ?*)

© Springer Nature Switzerland AG 2019
H. G. M. Edwards, *Porcelain to Silica Bricks*,
https://doi.org/10.1007/978-3-030-10573-0

Body Number 4:
Common porcelain body
12 FO composition china stone—4cwt 70lb; 8 YX bone ash—5 bone ash; 8 B St. Stephen's china clay—3 ½; 1 E Norden blue clay—35 lb.
Note: No sand used. The quantities of paste mixture specified seem large for a trial experiment—some 518 lbs of china stone alone (equivalent to 234 kilos).
(According to Eccles & Rackham, this is the "first evolution" from a duck-egg porcelain body).

Body Number 5:
Biscuit recipe first used in autumn 1815
20 parts V sand + 1 part FX pearl ash in water fritted at very high heat.
140 lbs of above frit + 110 FO china stone + 25 SR soap rock fired very regularly and gradually or it will blister. It was afterwards found that the blistering proceeds from an accidental mixture of alabaster to prevent the possibility of which great care must be taken.
(Believed to be an early effort at a *trident* soapstone body).

Body Number 6:
Variation on the above formulation. Sounder body but the articles still continue to "fly with hot water".
140 lb frit + 110 lb FO china stone + 35 lb SR soap rock
It was discovered that the B St. Stephen's clay and FO china stone when fritted together into one mass with FX pearl ash make an equally good looking body which stands well.

Body Number 7:
45 FO china stone composition + 10 LO lime + 28 B St. Stephen's china clay
Makes a body which comes very near the Chinese eggshell and will take a hard glaze but must be fired at a very high temperature and pieces are then apt to get out of shape.

Body Number 8:
Very good
9 parts V sand + 1 part B St. Stephen's clay and a little LO lime fritted in a very high heat.
3 of the above frit + 3 FO china stone + 1/10 SR soap rock.

Body Number 9:
B St. Stephen's clay of which half has been fritted and ground, glazed with FO china stone, is the *Dresden china*. A very great heat is necessary and difficult to get saggars to stand it. Equal parts of B St. Stephen's clay and FO china stone is the very best French china and will take an FO china stone glaze. No other than an FO china stone glaze will do as all others craze.

Body Number 10:
A beautiful china which stands well but is rather too soft for the hard glaze.
12 B St. Stephen's china clay + 12 YX bone ash + 12 FO china stone + 3 LO lime
(According to Eccles & Rackham, page 15, this approximates very closely to the
Nantgarw body recipe). No glass content.
Autumn 1816

Body Number 11:
A beautiful body and in all respects answers.
3 B St. Stephen's clay + 3 FO china stone + 3 YX bone ash.

Body Number 12:
An improvement.
8 B St. Stephen's china clay + 7 FO china stone + 8 YX bone ash
(According to Eccles & Rackham this is a *second evolution* on the duck's egg
body).

Body Number 13:
Makes the body harder but large pieces are more apt to fly.
9 B St. Stephen's clay + 9 YX bone ash + 7 FO china stone
This body glazes well with glaze number 2.
March 1817

Body Number 14:
12 V sand + 10 FO china stone + 2 FX pearl ash fritted together then 14 of this
frit + 2 SR soap rock makes a beautiful and good body. If only 1 SR soap sock is
used it makes the body whiter but the clay is more difficult to work. Afterwards, the
following alteration was made. (A second attempt at the *trident* body).

Body Number 15:
Without much improvement on the above body
8 V sand + 6 FO china stone + 1 FX pearl ash fritted in a high heat which had
better exceed the biscuit heat.
This body glazes well with glaze number 1.
December 1817

Body Number 16:
Makes a beautiful white opaque body and with glaze number 3 is the finest
earthenware I ever saw.
24 YX bone ash + 8 KO flint + 16 B St. Stephen's china clay + 5 E Norden
clay + 1 LX smalt.

Glaze Number 1:
Frit: 10 FO china stone + 6 LO lime + 2 B St. Stephen's clay + 12 V sand + 14
AX lead + 8 MX borax calcined + 4 EX nitre: total 56 parts
Run in the glaze kiln or earthenware biscuit kiln which is about the same heat I
prefer the latter on account of its longer continuance, which makes the frit run more
thoroughly throughout.

Then 56 parts of the above frit + 30 FO china stone + 6 LO lime + 2 B St. Stephen's clay + 14 AX lead + ½ GX arsenic

Glaze Number 2:
24 V sand + 12 LO lime + 6 AX lead + 16 MX calcined borax + 2 FX pearl ash Run in glaze heat or as Number 1
then 28 parts of the above frit + 40 FO china stone + 28 AX lead + 6 LO lime + 4 B St. Stephen's clay.

Glaze Number 3:
24 V sand + 12 LO lime + 6 AX lead + 16 MX borax calcined + 2 FX pearl ash frit in glaze heat as Number 1; then 48 FO china stone + 6 LO lime + 4 B china clay + 30 AX lead + 40 above frit + ½ arsenic to be dipped thick.

B.1 Notes

From Dillwyn's diaries it is believed that he commenced these experiments with Walker in August, 1815 and that they ran on until December, 1817, which is dated for his formulation *Number* 16. It is a commonly held view that Billingsley and Dillwyn were in disagreement about these trials designed to produce a more robust version of the esteemed duck-egg china and that this was achieved by a reduction in bone ash and increased content of soaprock. The resulting trident paste was truly harder but suffered in a decreased translucency, a muddier appearance and a dirty brownish-yellow light transmission which, with a less effective glaze, gave a rougher surface texture and a pitted appearance. The clientele immediately voted against the purchase of this inferior porcelain, even though it was still superbly decorated and evidently much more robust for handling and cleaning in boiling water. The adoption of eth trident paste into production is often cited as the reason for Billingsley and Dillwyn parting company in late 1816 as the trials were still ongoing: Billingsley left Swansea in December, 1816, but Walker stayed on to assist Dillwyn in the completion of his trials. Nevertheless, Walker left Swansea in September 1817, to join Billingsley and Young at Nantgarw to start up the phase II operation there, but Dillwyn still proceeded with his trials as the entry for number 16 in December, 1817, testifies. The production at Nantgarw in this Billingsley, Walker and Young era was at its highest acclaim and almost all the factory output was taken by Mortlock's of Oxford Street, London.

Appendix C
The Purple of Cassius

The purple preparation of colloidal gold and stannous hydroxide, tin (II) hydroxide, known historically as the "*Purple of Cassius*" has been consistently described in the literature as the invention of Andreas Cassius of Leiden in 1685 (*De Auro*, 1685), who added stannous hydroxide, $Sn(OH)_2$, to chloroauric acid, $HAuCl_3$, but it seems that this preparation had been in use for some years earlier and was well known to impart a rich ruby red colour to glass in the German glass factories in Potsdam (Hunt 1976). The latter property of gold when heated with alkaline silicates in the manufacture of red glass had been known for many centuries and had already been described by Agricola, Paracelsus, Glauber and Cellini. However, it was Glauber (*Des Teutschlandts Wohlfahrt*, 1656–1661) who first published the result of the reaction following the dissolution of tin in a solution of gold in *aqua regia*, HNO_3/HCl, which produced a deep purple precipitate of gold powder. The adoption of this purple gold pigment in glass manufacture dates from the observations of Kunckel (*Ars Vitraria Experimentalis*, 1679) which for many years afterwards became the staple handbook for ruby glass manufacture and colouring backed up by Orschall's *Sol Sine Veste*, published a year before that of Cassius in 1684. The preparation of the tin-based reduction of a solution of gold to its deep purple-red pigment was very much an alchemical art as described in several of these texts and was not easy to accomplish in practice, the precise recipe and process being surrounded by much secrecy and mysticism.

The idea soon spread to the decoration of ceramics and in 1719 the *Purple of Cassius* as it came to be known was being used as a pigment to decorate Meissen porcelain: an interesting reverse precursor of the much-quoted secret of Chinese porcelain composition which was first brought to France by Jesuit missionaries from Peking in the late 1720s, was the conveyance of the knowledge of this new and desirable purple pigment which was taken by the French Jesuits travelling to China—resulting in its use in the decoration of the famous Chinese *famille rose* porcelain by 1723 (Garner, *The Origins of Famille Rose*, 1967/9).

© Springer Nature Switzerland AG 2019
H. G. M. Edwards, *Porcelain to Silica Bricks*,
https://doi.org/10.1007/978-3-030-10573-0

   The identification of this pigment on Nantgarw porcelain by a combination of Raman spectroscopy and SEM/EDAXS techniques is a novel discovery: the use of gold in such a context has not been suspected hitherto and is indicative of the high quality and expensive enamels that Billingsley used in his ceramic art work. Gilding has been described by many previous authors as the most expensive process involved in the creation of decorated porcelains—most historians have assumed hitherto that this naturally refers to the application of gold leaf with an organic adhesive (such as *honey gilding*) or alternatively the use of a mercury/gold amalgam to the profusely decorated and gilded Regency porcelain pieces but the adoption of *Purple of Cassius* enamel pigment affords a further use for gold leaf in this context. Because Young and Pardoe decided to economise on their use of gold leaf for gilding, it is therefore extremely unlikely that they would have used gold to prepare a purple enamel in this way, and it can be confidently assigned to the Billingsley/Walker era at the Nantgarw China Works, a conclusion supported by its discovery on a phosphatic bodied and glazed porcelain shard. The beautiful colour tone and depth of deep purple colour conferred upon the Nantgarw translucent porcelain body by use of this synthetic pigment is truly admirable and Billingsley and Walker must be admired for their use of this pigment in what appears to be "no-expense spared" enamelling process.

References

A. Cassius, *De Extremo Illo et PerfectissimoNaturae Opificio ac PrincipeTerrae Nonum Sidere Auro* (Georgii Wolffi, Hamburg, 1685)
Sir H. Garner, The origins of *famille rose*, *Trans. Oriental Ceramic Soc.*, **37**, 1–16 (1967/9)
J.R. Glauber, *Des Teutschlandts Wohlfahrt*, Amsterdam, Part IV (1656–1661), pp. 35–36
F. Habashi, Purple of Cassius: Nano gold or colloidal gold?, *Euro. Chem. Bulletin.* **5**, 416–419 (2016)
L.B. Hunt, The true story of purple of Cassius: The birth of gold-based glass and enamel colours, *Gold Bulletin.* 9, pp.134-139, 1976.
J.C. Orschall, *Sol Sine Vest,* (Augsburg, 1684) pp. 18–19

Appendix D
Raman Spectroscopy of Porcelain Shards and Perfect Specimens

The molecular spectroscopic interrogation of Nantgarw porcelain shards and perfect, finished specimens has been undertaken for the first time by Edwards and Colomban (2018) and a full report is being prepared for presentation in the scientific peer-reviewed literature. Here, following a brief description of the Raman spectroscopic technique, the essential information emerging from the analytical experiments carried out on Nantgarw porcelain will be summarised and correlated with the combinatorial SEM/EDAXS elemental oxide determinations and the conclusions compared with existing historical opinion and knowledge in a holistic forensic approach.

D.1 Analytical Raman Spectroscopy

Raman spectroscopy is a molecular scattering technique, whereby a monochromatic laser beam is imaged onto the surface of a specimen and the back-scattered Raman radiation is collected, spectrally dispersed on a holographic grating and detected on a charge-coupled detector (CCD) device. The Raman spectrum is displayed as a series of wavenumber shifts from the laser generated Rayleigh line which acts as a wavenumber zero on the wavenumber shift scale (Long, *The Raman Effect: A Unified Treatment of the Theory of the Raman Scattering of Molecules* 2002). Being a molecular technique, the Raman Effect gives a unique spectral signature for materials which have chemical bonds, inorganic and organic molecules and molecular ions, and therefore complements other analytical elemental detection techniques such as XRD (X-Ray Diffraction), XRF (X-Ray Fluorescence) and SEM/EDAXS (Scanning Electron Microscopy/Energy Dispersive X-Ray Spectroscopy). It is especially valuable for the detection of molecular and molecular-ion components in mixtures

© Springer Nature Switzerland AG 2019
H. G. M. Edwards, *Porcelain to Silica Bricks*,
https://doi.org/10.1007/978-3-030-10573-0

and for changes in materials affected during chemical processing, as a result of procedures such as thermal reactions in ceramic materials. As far as ceramics and porcelains are concerned, the presence of a transparent glaze has no untoward adverse effect since the laser beam can penetrate the glazed layer and interrogate the underlying ceramic body; during this experiment, no damage is done to the sample by the imaged beam of light and no chemical or mechanical pre-treatment of the sample is required, such as removal of the glaze or of the surface polishing. The presence of individual minerals or materials is recognised from their observed spectral band wavenumber positions, which can be identified from literature databases. The intensity of Raman scattering is approximately proportional to the individual species concentration, so the greater the concentration of a particular material present the more intense the Raman band observed: not all materials have the same Raman molecular scattering factor cross-section for laser irradiation, however, so several species are always relatively more strongly represented in Raman spectra even when occurring in low concentrations, such as cinnabar, calcite, anatase and gypsum (Edwards 2015).

Although the primary investigative instrumental techniques for porcelains and ceramics that have been adopted thus far have been those where the elemental determinations have been paramount, such as XRay Diffraction (XRD), XRay Fluorescence (XRF) and Scanning Electron Microscopy (SEM), the advantage of a molecular technique such as Raman Spectroscopy (RS) is that it readily associates the combination of the chemical elements and their formulation through chemical bond identification: hence, although SEM can correctly provide evidence of mercury, lead, sulfur, oxygen and tin in an artwork or archaeological specimen it is often a matter of conjecture or expert opinion as to how these elements are paired together—for example, potentially these elemental data could signify the presence of the pigments cinnabar or vermilion (mercury sulfide), mosaic gold (tin sulfide), lead oxide (massicot, litharge, plattnerite and minium), lead sulfide (galena), calomel (mercury oxide), cassiterite (tin oxide), and Naples yellow (lead tin sulfide). The larger the number of heavier metals and anionic entities that are found in the material under study, then the more real possibilities are created for their pairing or association: sometimes, the colour of the particles under investigation will assist in a narrowing down of the possibilities, for example, a red pigment might be indicative of cinnabar or litharge, whereas a yellow pigment might be indicative of massicot, goethite, orpiment, realgar or mosaic gold (Edwards 2015). In all of these cases, the Raman spectral signatures are different and discriminatory and can assist in the evaluation of the correct pairing of metals and their anions. Even more problems can arise when one considers the elemental detection of calcium, magnesium, sulfur and oxygen: calcite, aragonite, gypsum, anhydrite, bassanite, magnesite, dolomite, epsomite and hydromagnesite are then all real mineral possibilities but unlike the pigment examples cited above, all of these are white, so the particle colour does not help at all in this case. Again, the Raman spectral signatures are discriminatory and can even identify mixtures of these materials, as in dolomitized calcite and in "hydrated lime" prepared from dolomite, which contains chemically calcium and magnesium oxides and hydroxides.

For ceramics, the presence of silica, silicates and sand, which all contain silicon-oxygen bonds, and Si=O or Si–O bonding combinations in three dimensional networks, give rise to silicaceous matrices which can be quite complex to describe when created at high temperatures. A description of the types of silicate network that can arise at the temperatures adopted for the kiln firing of soft-paste and hard-paste porcelains has been given in specialist texts and a summary has been provided by Edwards (*Swansea and Nantgarw Porcelains*: *A Scientific Reappraisal* 2017; Edwards, *Nantgarw and Swansea Porcelain*: *An Analytical Perspective* 2018). As well as the basic orthosilicates containing the discrete SiO_4^{4-} ions with four-valent silicon from the parent silicic acid, H_4SiO_4, condensed ions such as pyrosilicates and metasilicates also occur and network formation through the sharing of Si=O bonds with neighbouring silicon atoms gives rise to –O–Si–O– linkages in extended three-dimensional structures, conferring strength and rigidity upon the fired porcelain articles. All of these will have different Raman signatures, but the resultant spectral band response will be rather broadened by the number of alternative possible structural conformations of Si=O and Si–O bonding in clays and glasses, rather than discrete sharp features as expected for other more crystalline minerals (Colomban et al. 2001, 2004a, b, 2006). A major advantage of RS for porcelain analysis is the ability to interrogate the glazed body of a finished article through the glaze, which is transparent to the laser beam, and this opens up the specimen sampling procedure to include perfect examples of rare and museum pieces without the potential of incurring damage, unless the incident laser irradiance (power per unit area, W cm^{-2}) is very large or absorption of the laser radiation wavelength by a pigmented specimen causes localised laser heating. Hence, for the assessment of analytical information obtained from perfect finished porcelain, Raman spectroscopy could provide a viable solution for the requirement of a non-contact, non-destructive analytical procedure and it is with this in mind that the preliminary analyses of the Nantgarw porcelain specimens detailed above have been undertaken—the first on record for Welsh porcelain using this technique, although there are several key papers in the scientific literature which describe the non-destructive Raman spectroscopic analyses of porcelains: these include Medici porcelains (Colomban et al. 2004a), early European porcelains such as St. Cloud, Chantilly, Mennecy and Vincennes (Colomban et al. 2004b), Meissen (Colomban et al. 2006), Ming porcelain (Carter et al. 2017) and Rockingham (Edwards et al. 2007). Regarding the pigment analyses, there is an extensive database now in existence for the characterisation of enamels and pigments used in artworks generally, and ceramics specifically: for manuscripts and paintings (Bell et al. 1997; Burgio and Clark 1997), for enamels and glazes on ceramics (Colomban et al. 2001; Kock and de Waal 2007; Edwards et al. 2007), and a database for early European porcelain identification (Ricciardi et al. 2009).

The Raman spectroscopic analysis of porcelains undoubtedly and primarily addresses the identification of the mineral molecular composition of the bodies and glazes, and some key Raman spectral features of relevant materials are presented with their band wavenumbers in Table D.1, but secondary information can be acquired about the production processes and in particular the temperatures achieved

Table D.1 Raman Spectral Signatures of Minerals Relevant to Porcelain Bodies and Glazes

Mineral	Chemical formula	Key spectral wavenumbers/cm^{-1}
Albite	NaAlSi$_3$O$_8$	230, 285, 320, 408, 485, 510
Anatase	TiO$_2$	143, 200, 398, 517, 638
Anhydrite	CaSO$_4$	715, 1017, 1133
Anorthite	Ca$_2$Al$_2$Si$_2$O$_8$	195, 480, 510, 550, 980
a-Wollastonite	CaSiO$_3$	575, 988
b-Wollastonite	CaSiO$_3$	636, 972
Borax	Na$_2$B$_4$O$_7$·10H$_2$O	1400
Bytownite	CaNaAl$_4$Si$_4$O$_8$	226, 415, 468, 623, 975, 994, 1060
Carbon	C	1320, 1585
Cerussite	PbCO$_3$	152, 1052, 1370, 1477
Coesite	SiO$_2$	520
Cristobalite	SiO$_2$	230, 410, 520
Diopside	MgSiO$_3$	660, 1015
Enstatite	MgSiO$_3$	350, 405, 675, 1015, 1036, 1088
Fayalite	Mg$_2$SiO$_4$	820, 850
Feldspar	KAlSi$_3$O$_8$	190, 275, 325, 411, 488, 508, 522, 645, 973
Forsterite	Fe$_2$SiO$_4$	835, 860
Gypsum	CaSO$_4$.2H$_2$O	412, 481, 617, 1007, 1130
Haematite	Fe$_2$O$_3$	229, 299, 409, 640, 1320
Ilmenite	FeTiO$_3$	222, 371, 680
Leucite	KAlSi$_2$O$_6$	490, 515
Mullite	Al$_2$SiO$_5$	480, 600, 960, 1130
o-Silicate	SiO$_4^-$	450–460, ~1000, 1155
Quartz	SiO$_2$	205, 464
Rutile	TiO$_2$	144, 233, 445, 610
Sanidine	KNaAlSi$_3$O$_8$	170, 280, 475, 510
Serpentine/Chrysotile	Mg$_3$Si$_2$O$_5$(OH)$_4$	230, 345, 389, 620, 690, 1105
Tridymite	SiO$_2$	204, 310, 365, 420, 785
Whitlockite	Ca$_3$(PO$_4$)$_2$	960

in the kilns in the final stages of porcelain manufacture. Earlier work from Philippe Colomban and his team at the Universite de Pierre et Marie Curie in Paris has established a sound basis for the interpretation of the Raman spectra of intricate and highly complex silicates in terms of both the types of silicon-oxygen bonding present as well as the degrees of polymerisation of the silicate matrices which have been subjected to elevated temperatures in kiln fired porcelains, which have been categorised into five different types, namely Q^0–Q^4, varying from monomeric through to polymeric and sheet silicates. (Colomban et al. 2001, 2004a, b, 2005; Colomban 2013). In glasses and glazes found on finished porcelains, an estimate of the degree of polymerisation (DP) of the silicate matrix can be evaluated from the relative band intensities of broad features centred at 500 and 1000 cm^{-1}. Generally, this DP is measured as the $A_{500}:A_{1000}$ band intensity ratio which increases with the temperature to which the glaze has been subjected: hence, for porcelains fired at low kiln temperatures, which may typically be around 900 °C, the $A_{500}:A_{1000}$ ratio is in the range between 0.8 and 1.1, whereas for porcelains fired at kiln temperatures of up to 1400 °C this ratio rises to between 1.5 and 1.7. In the current study an important procedure is, therefore, the interrogation of the porcelain bodies of shard specimens analytically through the glaze coating: also, it is important to determine the effect on the resultant Raman spectra of the selection of the laser wavelength of excitation. Normally, the visible laser wavelengths 488 and 532 nm are used, but a laser wavelength used in several of our studies is in the near infrared region of the electromagnetic spectrum at 785 nm which tends to produce luminescence emission bands arising from the lanthanide elements which occur in china clays and which appear in the same region as the Raman spectral features.

The Raman spectroscopic molecular analysis of Nantgarw porcelains was adopted by the author and his collaborators to verify some key objectives:

Can one obtain definitive and key diagnostic Raman spectroscopic signatures for Nantgarw porcelain specimens which have been glazed and decorated?

Can one assign the observed Raman bands to the known components of the respective recipes after firing in the kilns and is it possible to correlate the bands observed with the absence or presence of components that may have been specified in the paste recipe, *viz*. the putative Nantgarw formulation and recipe published by John Taylor (*The Complete Practical Potter* 1847) which he allegedly obtained from Samuel Walker before he departed for North America?

Does the presence of the glaze interfere with the interpretation of the Raman bands or seriously mask their presence in the interrogation of the Nantgarw porcelain body underneath the glaze both in shards and perhaps, therefore, in perfect finished specimens?

Is it possible to discriminate non-destructively between genuine Nantgarw porcelain, pieces of dubious attribution and potentially deliberate fakes? This art forensic aspect of the scientific analysis of porcelains is an important input to expert opinion based on shape, texture and decoration to provide a holistic assessment of the factory origin.

Was there only a single Nantgarw body composition as has been maintained, perhaps naively and incorrectly, by many authors historically, but disregarding documentation that seems to suggest otherwise?

Is there a potential for the detection of "outliers" or associated pieces in large Nantgarw services, which may have been procured as replacements for broken or damaged items or which have been supplied originally to complete a commission order?

With these objectives and questions to be addressed, the interpretation of the Raman spectral data acquired from the Nantgarw China Works shard specimens held by the Nantgarw China Works Museum in Tyla Gwyn, Nantgarw, and described here becomes critically important for the establishment of the technique as a non-destructive analytical probe for the verification of perfect, complete and undamaged specimens which have been selected from an established collection of Nantgarw porcelain.

D. 2 Application of Raman Spectroscopy to Perfect Specimens of Nantgarw Porcelain

Noninvasive Raman spectroscopy has confirmed that the spectra obtained from the porcelain bodies of the glazed and unglazed Nantgarw shards were identical, which demonstrated that the surface glaze did not interfere with the data received from the subsurface porcelain body by the penetration of the interrogating laser beam radiation.

The spectra from the porcelain bodies of the glazed and unglazed Nantgarw phosphatic shards were identical with those obtained non-invasively from nine perfect, finished and decorated Nantgarw porcelain pieces presented for analysis. These specimens comprised: a dinner plate from the *Lady Seaton* service with impressed mark (Fig. D.1), a saucer from a London-decorated tea service painted by Moses Webster (Fig. D.2), a coffee cup from the *Spence-Thomas* service locally decorated by Thomas Pardoe (Fig. 4.1), an armorial dinner plate from the *Phippes* service with impressed mark (Fig. D.3), a London-decorated cylindrical spill vase (Fig. 1.3), a dessert plate from the *Sir John and Lady Williams* service (Fig. D.4), a dinner plate from the *Duke of Cambridge* service with impressed mark (Fig. 1.1) and two locally decorated dessert plates each with an impressed mark (Figs. 4.2a and D.5).

The Raman spectra using a stand-off probe and 785 nm excitation obtained non-invasively from five selected Nantgarw shards and the nine perfect specimens were distinct from those obtained from porcelain specimens originating from other factories, such as Swansea, Coalport, Derby, Pinxton and Worcester. This is an important result as it affords a potential method of discrimination between genuine Nantgarw porcelain and suspected attributions which could have originated elsewhere, especially porcelain that has been deliberately "faked" or made to simulate Nantgarw wares in the past.

Fig. D.1 Nantgarw porcelain dinner plate from the *Lady Seaton* service

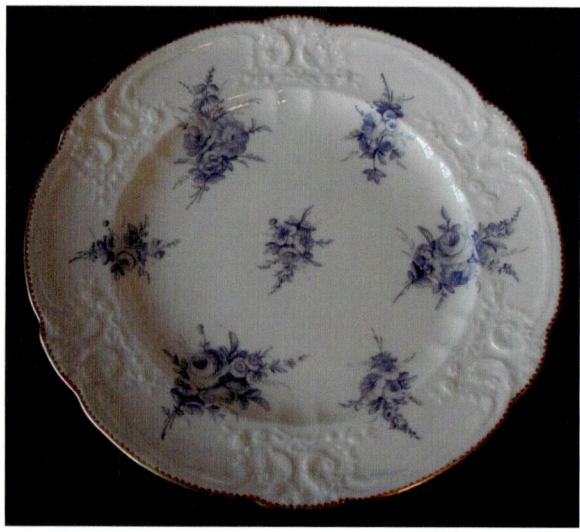

Fig. D.2 Nantgarw porcelain London-decorated coffee cup and saucer

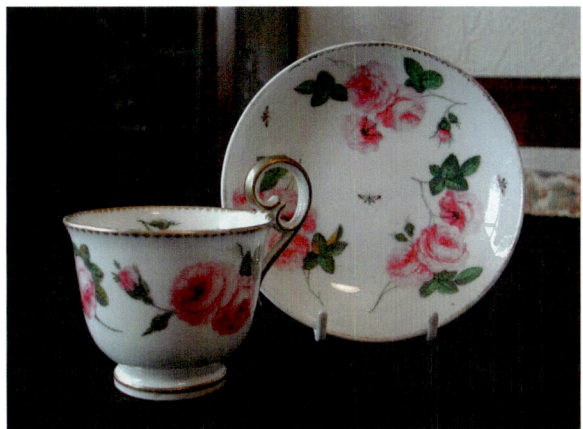

Fig. D.3 Armorial Nantgarw dinner plate from the *Phippes* service

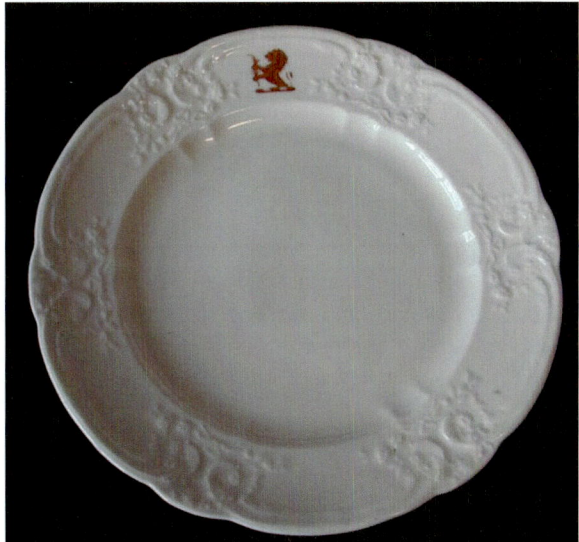

D.3 Blemishes in Nantgarw Porcelain

Attention has already been drawn to the appallingly high losses encountered in the firing of Nantgarw porcelain which accounted for up to 90% kilo losses through sagging and warping of the shapes. Prized for its fabulously high translucency as seen in Fig. 1.1, a rather unexpected blemish is sometimes seen on early porcelains and is shown here in Fig. 4.2b, c, which is the reverse side of a locally decorated dessert plate (Fig. 4.2a). This Nantgarw dessert plate, locally and simply decorated with sprays of orange roses, foliage and blue delphiniums and a simple edge gilding is marked with an impressed NANT-GARW C.W. The translucency is excellent but

Fig. D.4 Nantgarw porcelain dessert plate from the *Sir John and Lady Williams* service

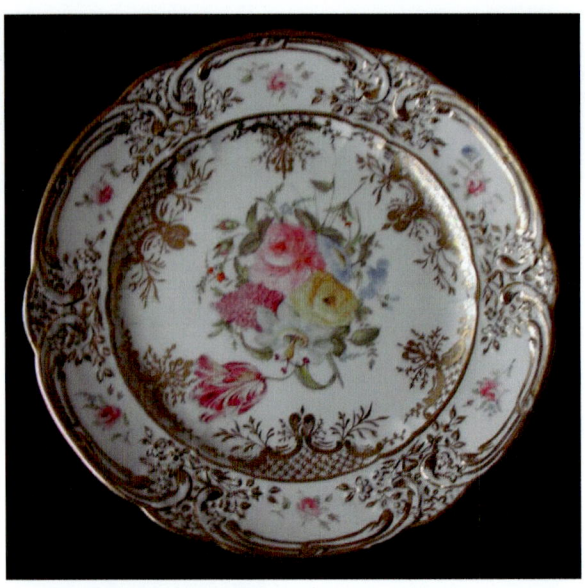

Fig. D.5 Nantgarw porcelain locally-decorated dessert plate, attributed to William Billingsley

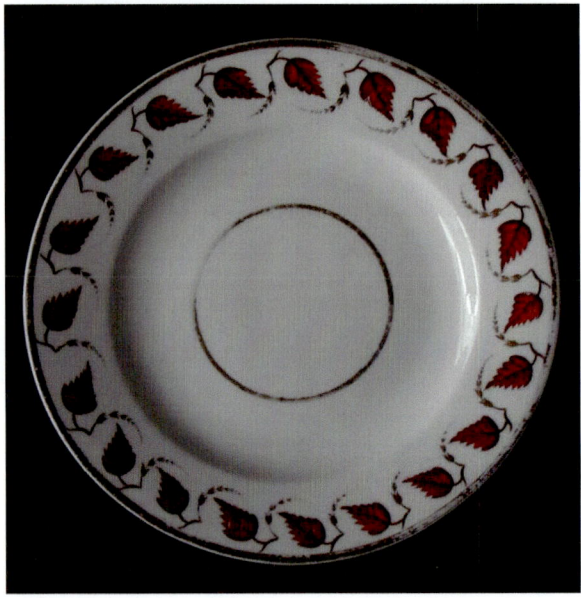

the presence of black spots and blemishes it is reasoned would have rendered this piece unacceptable for decoration in the London market through the agent Messrs. John Mortlock. Sometimes, small, dark blemishes or pits in an otherwise perfectly acceptable piece of porcelain could be masked by the strategic placement of insects such as butterflies and moths or by small flower buds: an example of this is seen in

Fig. 9.2, where several mosquitoes have been used to cover small blemishes the underside rim of the large Nantgarw meat platter from the *Farnley Hall* service shown in Figure 9.1 (Edwards, *Swansea and Nantgarw Porcelains: A Scientific Reappraisal* 2017). A Raman spectroscopic study of some of these dark blemishes on the locally decorated Nantgarw dessert plate has revealed two new, rather broad, features at approximately 1580 and 1320 cm^{-1}, which are characteristic of amorphous carbon, the so-called G (sp^2) and D (sp^3) bands, respectively, of hybridised carbon of graphitic or diamond-like structure. It can be concluded, therefore, that the black blemish on the surface of the Nantgarw dessert plate is elemental carbon.

It is interesting to speculate on the source of these blemishes: the Nantgarw kilns were oxidising in atmosphere (John, *Nantgarw Porcelain*, 1948) and it is perhaps surprising to think that finely divided particulate carbon could survive without conversion to gaseous carbon dioxide. However, inspection of the blemishes on the dessert plate analysed here reveals that they occur on the surface of the biscuit porcelain and have been covered by the glaze. This suggests that, following the removal of the porcelain after first firing and its treatment with liquidised slip for glost firing in the glazing kiln, some ingress of carbon particles became trapped in the wet glaze, covered by the glaze and became fixed thereupon for the second firing. The oxidisation of the entrapped carbon would thereby have been prevented by the hardened surface coating of glaze.

Occasionally, another sort of blemish is seen in Nantgarw porcelain such as that shown in transmitted light in Fig. 1.4, which occurs as a small, cream-coloured and rather diffuse isolated spot seen here near the factory impressed mark NANT-GARW C.W. on an armorial crested dinner plate from the *Phippes* service. This is entirely different in origin and can be attributed to either a stable mineral impurity in the paste or more probably to imperfect grinding and preparation of the powdered frit—leaving small pockets of feldspar or kaolin. The incorporation of some organic impurities, for example, would certainly give rise to blemishes of this sort and we can look no further than organic contaminants in the bone ash to be responsible for effects of this kind. A Raman spectrum could not be obtained from this blemish, which perhaps was situated too far beneath the surface for effective interrogation to be performed. Both Dillwyn and Billingsley were adamant in their selection of the purest possible sources for their calcined bone ash additive but occasionally some contamination could have permeated through the vetting processes. Such minor and interspersed blemishes do not detract seriously from the generally excellent translucency expected for Nantgarw porcelains and they are certainly of a different order of magnitude for visual impact to those attributed to the particulate carbon discussed earlier. In a contrasting viewpoint, it would also be reasonable to suspect that most organic contaminants would burn off at the high kiln temperatures necessary for porcelain production so maybe the source of this blemish arises from mineral contaminants as suggested.

References

I.M. Bell, R.J.H. Clark, P.J. Gibbs, Raman spectroscopic library of natural and synthetic pigments (pre-1850 AD). *Spectrochimi Acta Part A*. **53**, 2159–2179 (1997)

L. Burgio, R.J.H. Clark,Library of Fourier transform raman spectra of pigments, Minerals, pigment media and varnishes: A supplement to the existing library of Raman spectra of pigments with visible excitation. *Spectrochimica Acta Part A*.**57**, 1491–1521 (2001)

E.A. Carter, M.L. Wood, D. de Waal, H.G.M. Edwards, Porcelain shards from Portuguese wrecks: Raman spectroscopic analysis of marine archaeological ceramics. *Heritage Sci*.**5**, 17 (2017). https://doi.org/10.1186/s40494-017-0130-9

P. Colomban, B. Sagon, X. Faurel, Differentiation of antique ceramics from the Raman spectra of their coloured glazes and painting. *J. Raman Spectroscopy*. **32**, 351–360 (2001)

P. Colomban, V. Milande, H. Lucas, On-site Raman analysis of Medici porcelain. *J. Raman Spectroscopy*. **35**, 68–72 (2004a)

P. Colomban, I. Robert, C. Roche, G. Sagon, V. Milande, Identification des porcelains tendres du 18eme Siecle par spectrometrie Raman: St. Cloud, Chantilly, Mennecy et Vincennes. *Revue d'Archaeometrie*, **28**, 153–167 (2004b)

P. Colomban, V. Milande, On-Site Raman analysis of the earliest known Meissen porcelain and stoneware. *J. Raman Spectroscopy*.**37**, 606–613 (2006)

P. Colomban, The destructive/non-destructive identification of enameled pottery, glass artifacts and associated pigments—a brief overview. *Arts*, **2**, 77–110 (2013)

M.T. Domenech—Carbo, H.G.M. Edwards, A. Domenech-Carbo, J.M. del Hoyo-Melendez, J. de la Cruz- Canizares, An authentication case study: Antonio Palomino *versus* Vicente Guillo paintings in the vaulted ceiling of the Sant Joan del Mercat Church (Valencia, Spain). *J. Raman Spectroscopy*43, 1250–1259 (2012)

H.G.M. Edwards, *Historical Pigments*. ed. by R. Meyers, Y. Ozaki. Encyclopaedia of Analytical Chemistry (Wiley, Chichester, UK, 2015)

H.G.M. Edwards, *Swansea and Nantgarw Porcelain: A Scientific Reappraisal* (Springer, Dordrecht, The Netherlands, 2017)

H.G.M. Edwards, *Nantgarw and Swansea Porcelain: An Analytical Perspective*, (Springer, Dordrecht, The Netherlands, 2018)

H.G.M. Edwards, P.Colomban, B. Bowden, Raman spectroscopic analysis of an English soft paste porcelain plaque—mounted table. *J. Raman Spectroscopy*.**35**, 656–661 (2004)

H.G.M. Edwards, P. Colomban, to be published (2019)

W.D. John, *Nantgarw Porcelain* (Ceramic Book Co., Newport, 1948)

L.D. Kock, D. de Waal, Raman studies of the underglaze blue pigment on ceramic artefacts of the Ming dynasty of unknown origins. *J. Raman Spectroscopy*.**38**, 1480–1487 (2007)

D.A. Long, *The Raman Effect: A Unified Treatment of the Theory of Raman Scattering by Molecules*, John Wiley and Sons, Chichester, 2002.

P. Ricciardi, P. Colomban, B. Fabbri, V. Milande, Towards the establishment of a Raman database of early European porcelain. *e-Preservation Sci*.**6**, 22–26

J. Taylor, *The Complete Practical Potter* (Shelton, Stoke-upon-Trent, 1847)

I.J. Williams, *The Nantgarw Pottery and its Products: An Examination of the Site* (The National Museum of Wales and the Press Board of the University of Wales, Cardiff, 1932)

Glossary

Alabaster A type of white semi-translucent gypsum, calcium sulfate dihydrate, which in cleaved or thin sections was used in place of glass or in ornaments and statuary. Care needs to be taken in some earlier literature since the term has been applied to include calcite, calcium carbonate, and even a silicaceous onyx. Aragonite, a polymorph of calcite, is also found naturally as semi-transparent crystals and can often be confused with alabaster, although it is chemically very different in composition.

Biscuit porcelain The unglazed product of the first high-temperature firing process in the manufacture of porcelain using a "biscuit kiln" which operated in the region of 1200–1400 °C. The resultant porcelain is of a pure creamy-white or ivory colour and texture which if blemish-free and perfectly shaped, was very highly prized by ceramic artists and modellers particularly for the construction of ornamental figurines. In the late 18th Century, the finest biscuit porcelain articles were commonly placed on dining tables for admiration and as conversation pieces. Generally, the biscuit porcelain after cooling from the kiln was painted, glazed and gilded—during which any small defects could often be masked by strategically placed enamelling.

Burnishing The gentle polishing of gilt decoration on a glazed ceramics surface to a highly polished reflective coating. Early gilding was accomplished using "honey gilding" whereby 24-carat gold leaf was applied to the surface in a medium of honey or resins such as gum arabic which was replaced in the late 18th Century by mercury gilding using an amalgam of mercury triturated with gold leaf. During the final firing of the ceramic piece in the kiln at low temperatures the organic carrier component or mercury was volatilised leaving a dull golden finish, which was then hand-polished using finely powdered jeweller's rouge (iron oxide) to a brilliant finish.

© Springer Nature Switzerland AG 2019
H. G. M. Edwards, *Porcelain to Silica Bricks*,
https://doi.org/10.1007/978-3-030-10573-0

Calcination The process of heating a raw material to high temperatures and often red heat, to destroy organic and volatile components which might otherwise promote the evolution of gaseous products that could generate voids or blemishes in a viscous porcelain body paste under the firing process in the kiln. For example, limestone breaks down thermally to lime, calcium oxide, between 650 and 750 °C and the gaseous carbon dioxide evolved can be trapped in the matrix as unsightly bubbles. Powdered bones were generally heated to destroy the keratotic organic components in the hydroxyapatite matrix and coordinated water: the carbon formed initially during the heating process was converted to carbon dioxide at higher temperatures to leave a pure white material. At lower temperatures, artists favoured the heating of ivory or bone to lower temperatures but stopped the process at the formation of carbon, to harvest the "ivory black" or "bone black" pigment which was much appreciated for its depth of colour.

Coefficient of expansion and contraction A parameter which is based upon the ratio of linear change in size of a ceramic material upon the increase or decrease of temperature usually expressed as a unit of length per degree. It is critically important for both porcelain ceramics and bricks as it results in shrinkage of the initial article upon firing and also upon its subsequent use in kiln construction where the integrity of the structure could be compromised by a flexing of the structure or movement during large ranges of temperature and through the heating and cooling stages of operational thermal cycles.

Digestion A first-stage chemical process during which the components of a ceramic material are rendered soluble in water through their dissolution thermal in strong acids or alkalis such as hydrofluoric acid, sulphuric acid and sodium hydroxide. This is usually necessary to convert components such as silica, phosphates and silicates into water-soluble salts. After dissolution, the acidic or basic solution is neutralised and then divided into aliquot parts for chemical analysis.

Electromagnetic spectrum This describes the range of wavelengths which define the characteristics of electromagnetic radiation from the high-energy low wavelength cosmic and X rays through the ultraviolet and then the visible spectrum from the violet to the red and into the infrared, thence through to the low-energy, high wavelength microwave radiation. Analytical techniques probe different ranges of the electromagnetic spectrum to interrogate the molecular and elemental compositions of specimens and derive quantitative and qualitative data which can be correlated with the materials used in the ceramic bodies, glazes and pigments.

Firebrick A generic name for a variety of bricks of different compositional characteristics but which strictly is retained for clay bricks that have been made using thermal heating in kilns rather than baked clay bricks made through a sun-dried process at lower temperatures. Because the term refers to clay-based bricks it should not strictly be applied to silica bricks which do not normally contain any clay component—indeed, if this is the case the brick integrity can be compromised! Nevertheless, usage over many years has seen the adoption of terms such as refractory silica firebrick and silicate firebrick, which have a very different chemical composition and are used for different building and constructional purposes from thermal kiln longevity to frost-resistant building foundations.

Frit The result of fine grinding applied to the components of a ceramic paste especially used to describe compositions of bone ash china clay, flint glass (cullet), and soaprock. The necessity of very fine grinding to produce a homogeneous mixture of components which may have very tangible differences in hardness was appreciated by the earliest porcelain manufacturers and a fine frit usually then required just the addition of a single component or water to effect a suitable mixture for firing in a biscuit kiln.

Fusibility The process of melting a ceramic material at elevated temperatures to produce a molten phase with no residual solid residues is affected by the presence of alkaline components which are called fluxes typically soda, lime, potash and borax. The achievement and stabilisation of this molten phase at high temperatures is essential for the structural integrity of the porcelain body as the appearance of stress cracks can be initiated at lower temperatures at interfacial domains between fusible and infusible materials. The translucency of the final fired biscuit porcelain body is critically dependent upon the achievement of a single molten phase at the firing temperature used.

Gilding The application of a gilt decoration to fired porcelain involving 24-carat gold leaf in a carrier such as mercury honey or an organic resin followed by burnishing, see under burnishing.

Glost kiln Used in the final stage of porcelain preparation after applying enamelled decoration and involving the application of an alkaline "slip" containing china clay soda, potash and a lead oxide component usually in the form of a powdered flint glass, which at lower temperatures will form a hard, transparent glaze coating. Occasionally a glost kiln was used for calcination of components at lower temperatures and for the drying of components in which variable amounts of water were found to occur. Initially, glazes were lead-based but in the first decade of the 19th Century when the toxicity of lead compounds was appreciated, reversion to a tin-based glaze was effected (as patented by John Rose at the Coalport China Works in 1820).

Minor Additives Porcelain recipes or formulations frequently listed small percentages or amounts of additives such as cobalt blue or smalt, which was added to remove vestigial traces of yellow colouration in the incipient china clays occurring from small amounts of iron oxides, or ochres. Small amounts of borax, sodium tetraborate decahydrate, to increase fusibility and the plasticity of the body at higher temperatures and arsenic oxide to assist in the generation of a uniform silicaceous phase.

Ore smelting The process by which metal compounds in minerals and rocks are converted into their base metals by the chemical and thermal reduction of their oxides such as iron from iron ore (iron oxide), oxidation of their sulfides, or by chemical reactions of their sulfates, nitrates and phosphates. Particularly relevant in this study are the reduction of haematite to iron using coke and the oxidation of copper sulfide using oxygen to produce copper.

Pigments In ceramics decoration these are coloured minerals which are thermally stable at temperatures in the glost kiln after application of the glaze slip and firing up to 600 °C. Mineral pigments are usually metal oxides sulfates and sulfides such as haematite, gypsum and orpiment. Care needs to be taken in the interpretation of old recipes as minerals were often confusingly assigned an incorrect nomenclature, such as *minium*, which is used to describe red lead, trilead tetroxide, and cinnabar, mercury sulfide.

Slaked lime The product of the calcination of limestone (calcium carbonate) or dolostone (dolomite or dolomitised limestone) which results in the gaseous evolution of carbon dioxide and residual lime (calcium oxide) and/or magnesia. Reaction of this product with water gives calcium and/or magnesium hydroxide the so-called slaked lime or limewash, used as a white base coat for buildings or limewash putty used as a filler and mortar. On reacting with moist air, the lime or magnesia gives the respective carbonate, namely calcite or dolomite, which forms as a hard surface skin so preventing further substrate reaction.

Translucency This is perhaps the greatest achievable asset of porcelain manufacture to which every manufacturer subscribed and hoped to attain in emulation of the Chinese "eggshell" wares which were imported in the mid-18th Century. It describes the transmission of visible radiation or light, through a solid object: at one end of the scale is glass, which is *transparent* and at the other is earthenware or brick, which are *opaque*. Porcelain is measured by its translucency, which is the clarity for transmission when viewed with background lighting. An intermediate descriptor is "semi-opaque", which was applied to some china during the mid-19th Century, which is rather indefinite and conveys little information about their true category.

Wet chemical analysis The earliest type of chemical analysis for ceramics subdivided into *qualitative* which determined the chemical composition of the specimens and *quantitative*, which determined the proportions and relative percentages of each in the specimen. The analytical determinations involved the complete chemical digestion of the ceramic specimen and therefore its destruction. For ceramics analysis, the determination of silica, alumina, magnesia, lime, soda, potash, iron oxide and phosphorus pentoxide were normally undertaken, which could then be related back to the raw materials used in the formulation.

Index

© Springer Nature Switzerland AG 2019
H. G. M. Edwards, *Porcelain to Silica Bricks*,
https://doi.org/10.1007/978-3-030-10573-0

Printed by Printforce, the Netherlands